Safeguarding the Smallest Passengers: A Guide to Car Restraints for Children

Jonny

Table of Contents

Chapter 1: Introduction and book outline

1.1   Background and problem statement

Road traffic injuries are the leading cause of death in children and young adults between 5 - 29 years old (World Health Organisation, 2018a). South African road accident death rates are 41 per 100,000 for children under 5 years and 24.5 per 100,000 for 5 - 14 year olds (Puvanachandra et al., 2020). Children are at increased risk of fatalities and serious injury due to differences in their body segment proportions affecting body kinetics in a vehicle accident (Falkmer & Gregersen, 2001). Children's heads are proportionally larger and heavier leading to a higher centre of gravity and head inertia loading (Figure 1) (Brolin et al., 2015). Children are spending more and more time in a vehicle resulting in an increased exposure to the risk of vehicle accidents (Javouhey et al., 2006). Serious injury and death can be reduced by 25% with the use of seatbelts for rear seat occupants, 45-50% for front seat occupants and 60% for children with the appropriate use of car restraint systems (CRS) (World Health Organisation, 2018a).

Figure 1: The proportional changes in body segments from newborn to adult.

Copyright Volvo Cars, this illustration is inspired by Figure 5 in the publication: Burdi AR, Huelke DF, Snyder RG, Lowrey GH. Infants and children in the adult world of automobile safety design: Pediatric and anatomical considerations for design of child restraints, J. Biomechanics, Vol. 2, 1968:267-280

A recent observational study in South Africa found that 87% of children under 14 years were unrestrained and only 7.8% were restrained in a CRS (Puvanachandra et al., 2020). Similarly an observational study conducted at a local tertiary hospital found that only 8% of the children entering the premises in a one week period were seated in a CRS and 87% of children involved in motor vehicle accidents over a one year period were unrestrained (Kling, 2011). Yet, there are international policies and standards, and national legislation governing the use of CRS for children.

There are 12 Global Road Safety Performance Targets that have been developed as part of the 2030 Agenda for Sustainable Development (World Health Organisation, 2017). To achieve the goal of close to 100% seatbelt or CRS use by 2030 the United Nations Member States are urged to reduce the proportion of unstrained occupants by at least 10% each year by increasing seatbelt and CRS use (World Health Organisation, 2017). To ensure road safety, the World Health Organisation (WHO) recommends that children are transported in the rear of the vehicle in a CRS until they exceed the manufacturers weight and height recommendations (World Health Organisation, 2015a). Children with disabilities might require special consideration when being transported due to their additional physical needs (Lindner, 2011).

The South African government developed the South African National Development Plan in order to fulfil their equity goal "all children with disabilities have access to quality education" by 2030 (National Planning Commission, 2012). This implies that CSHCN should have safe transport to and from their educational institutions. Although many standard CRS are available, these might be unsafe or inappropriate for CSHCN of all ages due to the lack of postural support (Baker et al., 2012; Falkmer & Gregersen, 2002), for which more specialised CRS might be needed.

Several specialised CRS are available globally, however their availability in South Africa is limited and comes at a high cost (Sitwell Technologies, 2017). Therefore, the persisting lack of accessible transport for CSHCN reduces educational opportunities, hence jeopardising the fulfilment of the South African National Development Plan equity goal. Available surveys on the use of CRS in South Africa indicate the high cost of acquiring a CRS as the main reason for non-compliance to the South African legislation (Arrive Alive, 2015; Nagel, 2016), however, no literature focusing on transportation needs of CSHCN is available. Appropriate and affordable CRS are required for CSHCN to ensure their safety during transportation. Therefore, more evidence-based research is warranted, to determine the prevalence of using CRS in this population and the challenges, such as availability and financial burden, faced by parents. Identifying the specific demands for specialised CRS will aid future research to provide design recommendations for a locally manufactured and affordable specialised CRS. Therefore, this Master book investigated the use of and need for specialised CRS within the South African context.

1.2   Research questions and aim of the study

This book, focusing on CSHCN enrolled in Cape Town public special needs schools, consisted of two studies and addressed the following research questions:

1) What are parents' experiences of transporting their CSHCN?
2) What are the specific postural support requirements of CSHCN?
3) Are the CRS currently available in South Africa suitable for the South African CSHCN population?

In response to these questions, these studies aim to:

I. investigate the need for a specialised seating system with adapted postural support in transport systems and
II. identify criteria, such as diagnoses and functional ability, associated with the need for a specialised CRS.

1.3   Justification for the study in South Africa

According to the World Bank, South Africa is a middle-income country experiencing income inequality as a large share of the income goes to a small proportion of the population (Leibbrandt, Finn & Woolard, 2012; *World Bank Country and Lending Groups*). The median monthly income of employees is only R3 500 (Statistics South Africa, 2017) as almost a third of South Africans are unemployed (27.5%). Resulting in many South Africans in the bottom half of the income distribution relying on social grants (Leibbrandt, Finn & Woolard, 2012), influencing their ability to self-fund private transport and the appropriate CRS.

There are only a small number of specialised CRS available in South Africa for CSHCN most are considerably expensive (Sitwell Technologies, 2017), resulting in few families who can afford to purchase a CRS without the support from government funded services. However, there is a large gap between South African government policies and their implementation resulting in inequalities for CSHCN (Saloojee et al., 2007). In developed countries, CRS specifically designed for CSHCN are available (Stout, Bull & Stroup, 1989), and are more accessible for parents as the market is more competitive, resulting in refined designs and a wider variety of sizes to make provision for the diversity of CSHCN (McIntosh, Lindner & Suratno, 2013).

Although surveys on the use of CRS for South African children indicate that the high cost of acquiring a CRS as the main reason for non-compliance to the South African legislation (Arrive Alive, 2015; Nagel, 2016), there is currently no empirical evidence to support these claims among the CSHCN population. Therefore, more evidence-based research is warranted, to determine the prevalence of using CRS in CSHCN and the challenges, such as availability and financial burden, faced by parents. These challenges may prevent parents from acquiring the most suitable car restraint for their child's needs, resulting in parents creating customised devices which do not meet standard specifications to ensure the safety of the child during transit (Falkmer & Gregersen, 2001). Hence, understanding the current postural support needs during transportation of South African CSHCN will help to identify the specific demands for specialised CRS, and therefore provide a framework for the specifications of an affordable, specialised CRS.

1.4   Research setting

There are more than 13 million South African learners enrolled in the basic education system of which, 117 477 CSHCN attend 447 special needs schools nationwide (Department of Basic Education, 2016). Public special schools are stratified by the level of intellectual disability and not the learners' physical disability (Foskett K, 2014). For this book, we therefore selected three institutions, one for each type of special school or centre (except autism/deaf schools), to obtain a more adequately representation of the population of CSHCN (Figure 2).

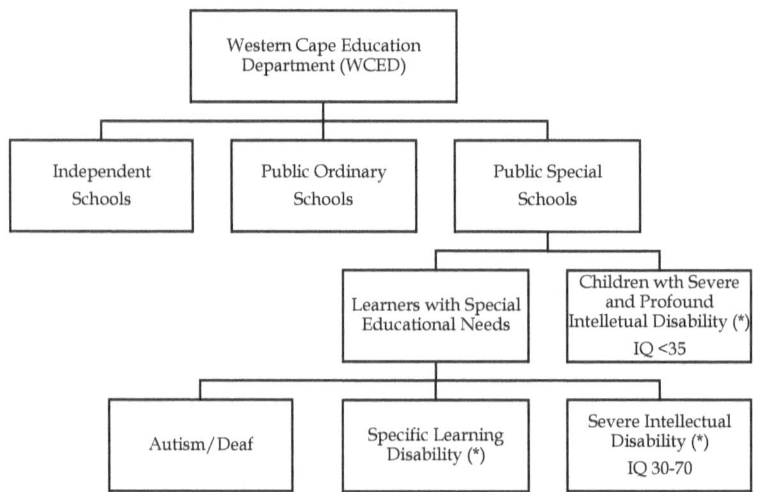

Figure 2: Organogram of Western Cape Education Department Schools with included school types indicated by (*).

Although CSHCN may be enrolled in special schools for deaf or autistic learners, or in independent schools or ordinary public schools as part of inclusive education (Department of Basic Education, 2015), these schools were excluded due to their low prevalence of CSHCN (Department of Basic Education, 2015). Three institutions, one of each type of public special school, were selected by convenience (Table 1).

Table 1: Public special schools selected for study

Type of School	Learners
Public special school for Special Learning Disability (SLD)	350 learners with specific learning disabilities age 4 to 18 years
Public special school for Severe Intellectual Disability (SID)	230 learners with severe intellectual disabilities age 4 to 18 years
Special care centre for Children with Severe and Profound Intellectual Disability (CSPID)	120 learners with profound and severe intellectual disabilities age 2 to 41 years

Common diagnoses at the selected schools include cerebral palsy, spina bifida, muscular dystrophy and Down syndrome. Learners at all three institutions may have mobility limitations, either walking with or without an assistive device, or mobilising in a manual or electrical wheelchair or have profound physical limitations.

To answer this book' research question, different chapters explaining various aspects of the research process have been created. The next chapter presents a general literature background to highlight the rationale for the research studies. Topics covered in that chapter will include the implication of physical disability on CRS, the national legislation governing the use of CRS for children and the international recommendations for CSHCN. Other research on this topic will be discussed which highlights the need for further research on CRS in the South African context.

Once the rationale has been established, chapter three will describe the process of testing the psychometric properties of the self-designed questionnaire used in a pilot survey study. As well as the survey study which was conducted at three schools to investigate the current transport systems in use and the challenges experienced by the parents. This provides an understanding of the current context for the CSHCN on their daily commute.

The study in chapter four tested the suitability of different CRS designs for CSHCN. It was done by firstly conducting a pilot study to establish inter- and intra-rater reliability of the screening tool designed by the researcher and to pilot the feasibility of the study. The study then compared the postural support offered by different CRS designs and a seatbelt for CSHCN by comparing postural deviations of the head and trunk once the CSHCN experienced forces similar to that of a moving vehicle. This provides an understanding of the physical support and restraint needs of a specialised CRS.

Finally, chapter five.summarises the results from both studies and the transportation needs of the CSHCN within the South African context.

Chapter 2: General Background

2.1 Introduction

The majority of the 150 million children with a disability worldwide, live in low- and middle-income countries (Maulik & Darmstadt, 2007). The estimated prevalence of childhood disability in South Africa is more than one million in a population of thirty-eight million (Saloojee et al., 2007; Statistics South Africa, 2014). Despite the implementation of the South African Schools Act, making education compulsory for all children aged 7 to 15 years (*South African Schools Act No. 84 of 1996*), only 117 477 of children with a disability attend the 447 special schools nationwide (Department of Basic Education, 2016). School attendance is highest amongst children with no disability however, more than a third of children with severe physical disabilities are not attending school (Statistics South Africa, 2014). Access to schools remains a challenge due to the lack of adapted transport accommodating the complex physical needs of these learners (Department of Basic Education, 2015).

CSHCN have been defined as, "those who have or are at increased risk for a chronic physical, developmental, behavioural, or emotional condition" (McPherson et al., 1998:138), this includes children which have motor, sensory and/or intellectual disabilities (Saloojee et al., 2007). A key problem in children with motor impairments is limited postural control that interferes with their activities of daily living (Brogren, Hadders-Algra & Forssberg, 1998; Field, D & Livingstone, 2013). The conceptual framework of the International classification of functioning, disability and heath – child and youth version (ICF-CY) provides a universal coding system to accurately describe child's disability and how it impacts their daily life (McDougall & Wright, 2009). The interactions between the components of the ICF-CY are illustrated in Figure 3. In particular, the environmental factor of transportation for CSHCN can be described by chapter 4 of the ICF-CY which details mobility, specifically moving around using transportation (World Health Organization, 2007). Children with motor impairments often use vehicular transport to travel to school, hospital or recreational activities and such transportation should be made safe and comfortable (Baker et al., 2012). To ensure CSHCN safety and comfort, there is a need to incorporate adaptive seating systems for improved function and sitting ability (Tecklin, 2008:209).

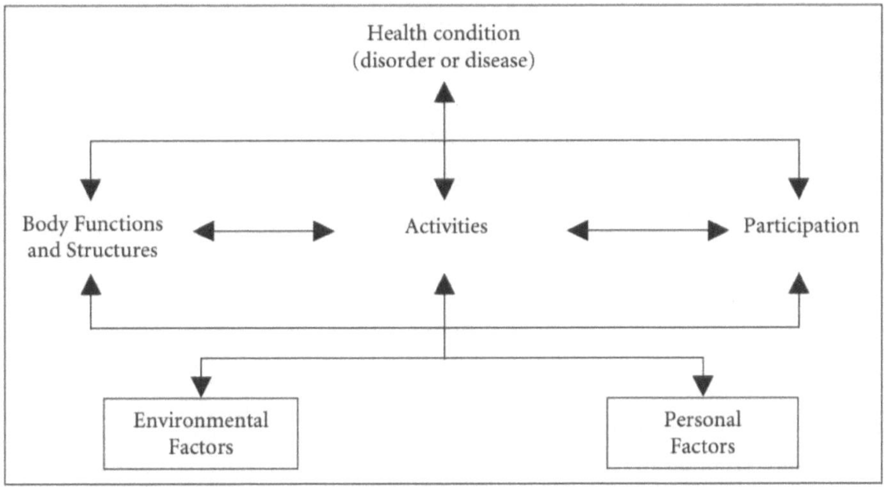

Figure 3: Interactions between the components of ICF (World Health Organization, 2007)

2.1.1 The implication of physical disability on seating for children

A recent policy statement issued by the American Academy of Pediatrics (AAP) recommends that, "all children, including those with special health care needs, should have access to proper resources for safe transportation" (O'Neil & Hoffman, 2019:1). Provision of transportation for CSHCN is complex, largely understudied and faces many barriers (Korn et al., 2007). These barriers include insufficient space to transfer in and out of the vehicle (Petzäll, 1995), travel time in the vehicle exceeding the CSHCN postural endurance (Department of Basic Education, 2015) and lack of adaptive seating for improved postural support and sitting ability within the vehicle.

Adaptive seating forms part of the therapeutic management to address poor postural control and could entail additional modification of these adaptive seating devices (Chung et al., 2008). These devices provide support at key points throughout the body, particularly the pelvis and trunk, limiting the degrees of freedom at each joint by making use of additional support mechanisms such as pommels, headrests or straps (Healy, Ramsey & Sexsmith, 1997). The adaptive seating device aims to improve positioning; performance of functional tasks requiring postural control and balance resulting in enhanced participation, and prevention of long-term complications, such as pressure sores, deformity and reduced function (Angsupaisal, Maathuis & Hadders-Algra, 2015; McDonald, Surtees & Wirz, 2004).

2.1.2   Transportation of children with special health care needs

Children with motor impairments often use vehicular transport to travel to school, hospital or recreational activities, which should be safe and comfortable (Baker et al., 2012). Provision of transportation for CSHCN is complex, largely understudied (Korn et al., 2007), and faces many barriers, including insufficient space to transfer in and out of the vehicle (Petzäll, 1995), lengthy travel time (Department of Basic Education, 2015) and lack of appropriate adaptive seating for improved postural support and sitting ability within the vehicle (Ryan & Rigby, 2007).

CRS are designed specifically to the anthropometry and postural needs of children, ensuring safety and proper fitment of the seatbelts (van Rooij et al., 2005). However, movements during non-crash events, such as swerving and braking, pose additional postural challenges for CSHCN (World Health Organisation, 2009), for which additional supports might be required (Andersson, Bohman & Osvalder, 2010).

Despite CSHCN meeting anthropometric requirements for standard CRS, those with physical, cognitive or other impairments may require specialised CRS (McIntosh, Lindner & Suratno, 2013). Those with a physical disability may require additional postural support in order to remain seated correctly (Lindner, 2011). CSHCN older than 14 years often require a larger CRS and additional support than standard devices or only a seatbelt (Lindner, 2011).

Physical conditions affecting muscle tone or posture such as hypotonia or scoliosis affect a CSHCN's ability to fit into a seatbelt or CRS (Huang et al., 2011). In particular, for CSHCN with orthopaedic conditions, breathing difficulties, reduced head and trunk control or changes in muscle tone, standard CRS may be inappropriate (Baker et al., 2012). For example, CSHCN with appliances such as orthoses or CSHCN with spasticity often struggle to use a standard CRS due to their limited joint range of movement (Baker et al., 2012). Medical conditions may require flexibility of the seat posture or adjustability of the harness (Lindner, 2011). However, these children also need to be positioned in a CRS as per international and national legislation.

2.2 Legislation, transportation services and the availability of restraints

2.2.1 Legislation & terminology

2.2.1.1 *Understanding seatbelts and car restraint systems*

The WHO categorise seatbelts and CRS as; "secondary safety devices that are primarily designed to prevent or minimise injury to a vehicle occupant when a crash has occurred by distributing the forces of a crash over the strongest parts of the body" (World Health Organisation, 2009:7). Seatbelts can include two-point lap belt and the three-point lap and diagonal seatbelt, whereas CRS have different groups for children of increasing mass. CRS are broadly grouped into categories based on the child's weight with specific design features based upon their physical needs. Infants under the age of 1 year should use a group 0 or 0+ infant seats which is designed for a child mass of less than 13kg, toddlers aged 1 – 4 years not exceeding 18kg should use group 1 Car Seats, children aged 4 – 6 years should use group 2 Car Seats provided they do not weigh more than 25kg and finally group 3 Booster Seats are for children older than 6 years and under 36kg (Figure 4) (World Health Organisation, 2009).

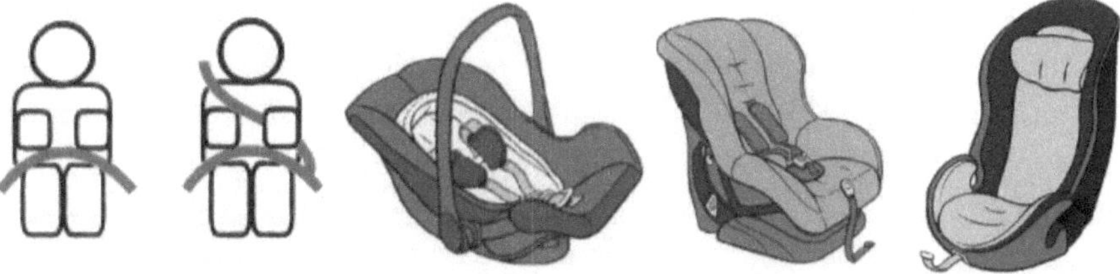

Figure 4: Examples of seatbelt and CRS types (from left to right: two-point lap belt, three-point lap and diagonal seatbelt, infant seat, car seat and the booster seat) (World Health Organisation, 2009)sin

CRS provide specialised protection for a child dependent on their size and weight (Charlton et al., 2010), and significantly reduces the risk of serious injury in an accident (Zaza et al., 2001). The different CRS designs are designed with the child's developmental age in mind and should be used until the seatbelt fits correctly (World Health Organisation, 2018b). Both weight and height, more so than age of the child should be considered when determining the type of CRS the child should use (World Health Organisation, 2018b). Children should remain seated in a rear-facing CRS for as long as possible for optimal protection (World Health Organisation, 2009). Progression from a CRS to a seatbelt is mainly determined by the child's height to ensure proper fitment. Misuse of the seatbelt, use of a two-point lap belt, or premature graduation from a CRS to a seatbelt are associated with risk of serious injury or death (Reeve et al., 2007).

The three-point lap and diagonal seatbelt is recommended by the WHO as the most effective design and it is the only belt that allows for the installation of a CRS (World Health Organisation, 2009). An alternative way of securing the CRS in the vehicle without the need for a seatbelt is via a standardised fitting system such as ISOFIX, LATCH or UAS (World Health Organisation, 2009). Since 2014, passenger vehicles are manufactured with these brackets mounted to the chassis of the vehicle providing a rigid, permanent connection between the CRS and vehicle (Newby, 2012).

2.2.1.2 *South African legislation on the use of seatbelts and child restraint systems*
The South African National Road Traffic Regulations with regards to the transportation of children (Department of Transport, 2014), is starting to align with international standards. According to Road Traffic Regulation 213, infants and young children, under the age of 3 years, must be seated in an appropriate CRS while being transported in a vehicle; and a child, up to the age of 14 years, must use an appropriate CRS where available (Department of Transport, 2014). The legislation on CRS use only applies to private transport and does not include public vehicles, such as minibus taxis or school buses transporting children daily (Department of Transport, 2014). According to the National Road Traffic Act children older than 14 years or taller than 1.5m are classified as an adult and are not required to use a car seat but must wear a seatbelt only (National Road Traffic Act, 1996). International best practice requires all children up to the age of 10 years or 135cm to use a CRS and there must also be restrictions of children sitting in the front seat of a vehicle (World Health Organisation, 2018a).

South African law stipulates that seatbelts and car restraints fitted into vehicles must comply with standard safety specifications SABS 1340 *"Child restraining devices in motor vehicles"* (Department of Transport, 2014). The National Road Traffic Act 93 of 1996 requires motor vehicles, first registered after September 2006, including passenger vehicles such as school buses, public transport services and minibuses, to be fitted with seatbelts. However, the belt configuration for the rear seats may be either a two-point belt or a three-point belt (National Road Traffic Act, 1996), resulting in the inability to use a CRS when travelling on public transport.

The Department of Education has identified that the lack of accessible transport is a key barrier for school attendance among CSHCN and hence, are developing the School Transport Policy (Department of Basic Education, 2015). This policy aims to standardise the implementation plans and alignment of strategic frameworks to deliver quality and reliable public transport to and from schools for CSHCN, but makes no mention of the use of CRS (Department of Transport, 2015). The current National Learner Transport Policy requires vehicles to meet the safety standards set out in the National Road Traffic Act of 1996 (Department of Transport, 2015). These policies should aim to align with the guidelines set out by the WHO for child safety, by ensuring that three-point seatbelts and CRS are made available to all children until they are anthropometrically fit for a seatbelt in all public and private vehicles.

2.2.2 Modes of transportation available in South Africa and the use of car restraint systems

2.2.2.1 *Modes of transportation in South Africa*

During a national survey it was found that only 26.1% of South African households have access to a car, with private vehicle ownership at 168 per 1000 in metropolitan areas which is lower than first world countries (Walters, 2013). Therefore, the majority of South Africans use public transport services including buses, taxis (mini-buses) and trains managed by the Department of Transport (Department of Transport, 2020). However, commuters reported many problems when using public transport, such as long distance to access public transport, unavailable transport, crowding on transport, concerns about safety and the high cost of transport (Walters, 2013). The literature on the accessibility of public transport, particularly for CSHCN, is lacking in the South African context.

Discussions held with parents in a European study indicated that public transport, including buses, was a poor alternative for CSHCN due to accessibility and reliability (McManus et al., 2006). According to this parent report, countries such as France have appropriate ramps for wheelchairs to access public transport, whereas nearly all Swedish parents felt that public transport was a problem and in Italy the school buses were not suitable for the disabled (McManus et al., 2006). Even if the vehicle is adapted it may not be user friendly, resulting in parents needing to accompany their child on every trip (McManus et al., 2006). As a result, nearly all Danish families report that the family car has been adapted for better access for CSHCN in a wheelchair (McManus et al., 2006).

Transportation via bus is a common mode of transport for the daily commute to school amongst studies, but the accessibility and safety for CSHCN's remain unclear (Downie et al., 2019; Falkmer & Gregersen, 2001; McManus et al., 2006). When CSHCN are appropriately restrained, travelling by bus may be safer than car because of the greater vehicle mass and slower travelling speeds, resulting in fewer accidents and reduced impact for CSHCN during a crash (Downie et al., 2019). However, Falkmer and Gregersen (2001) suggested that school transport was not a safe option due to a high proportion of CSHCN using inappropriate restraints or none at all, which they felt was a result of lax regulations on large capacity buses. Downie et al. (2019) recommended that further research investigate if the use of public transport systems is a safe option for CSHCN.

2.2.2.2 *Current seatbelt and CRS usage*

Promoting and improving parental seatbelt use can positively influence children's risk behaviour decisions concerning seatbelts (Morrongiello, Corbett & Bellissimo, 2008). A recent WHO Youth and Road Safety Report reported that seatbelt usage varies greatly between countries and is affected by the laws governing seatbelts to be fitted in cars and to be worn by users (World Health Organisation, 2018b). Rates of seatbelt use are lower in low-income countries if there are no laws requiring them to be fitted or used or if the laws are poorly enforced (World Health Organisation, 2018b). Unregulated taxi industries can also impact the ability to enforce the laws (World Health Organisation, 2018b). A local survey found that more than two thirds of urban passengers in the Western Cape were not wearing a seatbelt (Arrive Alive, 2013). This aligns with data from the WHO Global Health Observatory which has recorded South Africa's seatbelt usage at 31% (World Health Organisation, 2015b). Statistics distributed by the Child Accident Prevention Foundation of South Africa state that 84% of children do not use a seatbelt when travelling (Child Accident Prevention Foundation of Southern Africa, 2013). Furthermore, local observational studies of private vehicles have found that only 8-12% of children were using a CRS (Clay et al., 2019; Kling, 2011) with the most recent study reporting only 7.8% (Puvanachandra et al., 2020). This is in stark contrast to other countries such as Canada, New Zealand and Sweden where the rates of wearing a seatbelt is over 90% (World Health Organisation, 2015b). Unfortunately there has also been high frequencies of inappropriate CRS usage reported such as loose straps and harnesses or the CRS not secured to the seat (World Health Organisation, 2018b).

2.2.3 CRS recommendations and availability for CSHCN

Transportation needs of children with physical disabilities is complex and safety cannot always be guaranteed in standard CRS (Falkmer & Gregersen, 2001). Specialised CRS and standard CRS differ in adjustability, attachments, postural support, and usability (McIntosh, Lindner & Suratno, 2013). The body mass range is often much greater in specialised CRS to accommodate the older CSHCN that cannot be safely restrained by the seatbelt (McIntosh, Lindner & Suratno, 2013). Unfortunately, the field of CRS for CSHCN remains under-researched, even at a global level (Falkmer & Gregersen, 2002).

2.2.3.1 *Car restraint systems available internationally for CSHCN*

The need for specialised CRS was documented by Everly et al. (1993), particularly for CSHCN who no longer fit in standard CRS but cannot ride using only a seatbelt without additional support. Forty-four percent of parents of CSHCN in an USA based study were aware of the availability of a range of large specialised CRS and there were four different types of seats observed in use (O'Neil et al., 2009). Specialised CRS are similar to standard CRS in that they typically offer either three or five point harness systems, but differ in that they include adjustability, attachments, postural support, body mass range and usability (McIntosh, Lindner & Suratno, 2013). Various national standards bodies regulate the design and performance of CRS to guide their use and ensure the safety of CSHCN during a crash (Economic Commission for Europe of the United Nations, 2011; National Child Passenger Safety Board, 2012; O'Neil & Hoffman, 2019; Standards New Zealand, 2013).

Regrettably the misuse or inappropriate choic e of restraint of seatbelts, CRS or safety harnesses can jeopardise the safety of the children in the vehicle (Everly et al., 1993). Misuse of CRS includes any deviation from manufacturer's instructions, incorrect or inadequate anchoring of the CRS to the vehicle seat (Korn et al., 2007), as well as shortening, loosening or twisting of harness straps (Brown et al., 2010). Surveys and observational studies have shown a high prevalence of non-use or misuse of CRS for CSHCN (Korn et al., 2007; O'Neil et al., 2009) with resultant associated serious injury (Baker et al., 2012; Lindner, 2011).

The AAP recently issued two policy statements providing guidance for primary care providers, therapists and child passenger safety technicians for the safe transportation of CSHCN (O'Neil & Hoffman, 2019), and one specifically addressing school transportation

(O'Neil & Hoffman, 2018). These guidelines propose which CRS are most appropriate, how to determine proper fitment as well as guidelines for CSHCN with specific medical conditions (O'Neil & Hoffman, 2019). They recommend that CSHCN should be seated on the rear seat until at least 13 years old, and wherever possible in a standard CRS until they exceed the weight, height and/or length of the seat as recommended by the manufacturer (O'Neil & Hoffman, 2019). Older children who have exceeded the CRS limits may be safely transported with the seatbelt only, provided that the belt-positioning is appropriate and additional postural support is not required (O'Neil & Hoffman, 2019). It is recommended that school buses be fitted with three-point lap and diagonal seatbelts and other basic requirements for the safe transportation of CSHCN in a CRS or their wheelchair including, a platform lift, wheelchair tiedowns or additional anchors and tethers for the CRS (O'Neil & Hoffman, 2018).

It is difficult to apply these guidelines in the South African context because of the lack of accessible and affordable standard CRS and subsequent low usage (Puvanachandra et al., 2020). In addition, there is a high demand on the public transport system (Walters, 2013) and the vehicle regulations exclude the use of CRS in minibus, midibus or bus operating for reward (Department of Transport, 2014).

2.2.3.2   *Car restraint systems available in South Africa*
In South Africa, a small number of specialised CRS are available for CSHCN. Although, they provide multiple levels of postural support, customisation and ease of access; most are considerably expensive ranging between approximately R46 000 and R65 000 without accessories and have a weight restriction of 36kg (Sitwell Technologies, 2017). The cheapest option is R11 500 (Wheelwell, 2017), which is comparable in price to a locally produced postural support wheelchair (Shonaquip, 2017), however it is still more than three times the median monthly income (Statistics South Africa, 2017). There is no locally produced specialised CRS for CSHCN available to meet postural support needs in South Africa. There is however an in-vehicle posture support seat which must be used in conjunction with the vehicles safety belt system, but it has not been crash tested and thus not recommended as a safety seat (Shonaquip, 2017).

The majority of South Africans do not have private medical aid and thus depend on government funded services (Statistics South Africa, 2015). However, delivery rollout of these devices is hindered by budget restraints, insufficient supply and long waiting lists. This results in few families of CSHCN who can afford this necessary device for their child. Many families may have to compromise the amount of postural support the child receives while seated in a vehicle, or compromise meeting the regulatory safety standards. The increased risk of injury or death is significant for incorrect CRS usage, whether as a result of the CSHCN age or the physical needs (Teerds & Cameron, 2015).

2.2.3.3 *Alternative seating solutions*

Due to the high cost of specialised CRS and the need for additional postural support than that provided by standard CRS, parents have had to seek alternative seating solutions (Brinkey, Manary & Santioni, 2011). As witnessed at local special schools for CSHCN, current seating methods used by South African parents and caregivers of CSHCN vary immensely, depending on the severity of the disability and the resources available.

Observations include mechanical platforms at the rear of a large vehicle which is made to lift the child seated in the wheelchair into the vehicle. The wheelchair is then secured to the floor. Alternatively, locally available standard CRS or household manufactured car inserts are used, and in some cases CSHCN are even made to lie down on the rear seat without a restraint for those that cannot maintain their own balance. Wheelchairs and customised seating systems are not designed for in-vehicle usage and are not crash tested, increasing the safety risk for CSHCN users (Falkmer & Gregersen, 2001). For many parents mechanical wheelchair platforms and tiedowns are unaffordable and cumbersome, requiring permanent alterations to the vehicle (Easy Drive WC, 2016; Ryan & Rigby, 2007). Lying down against a door or being out of position may put a CSHCN at increased risk of injury should an accident occur (O'Neil & Hoffman, 2019).

Child safety harnesses, comprising of a lap strap attached to two straps over the shoulder, have not been shown to be more effective than standard seatbelts in a study on their safety (Brown et al., 2010). These harnesses might not meet the all the postural support requirements of CSHCN (Ryan & Rigby, 2007). Specialised harnesses usually require permanent fixture to the vehicle and thus cannot be transferred to a different vehicle. Informal observations in the South African context have shown that child safety harnesses do not provide sufficient postural support for CSHCN with complex seating needs, for which additional lateral trunk support, seat wedge or an abduction pommel may be required. Another alternative means of support could be provided by a child vest which provides limited trunk support to a CSHCN who can maintain head control (Stout, Bull & Stroup, 1989). These vests have a 5-point harness design with a universal attachment system to be used in school buses, vehicles and wheelchairs. However, it is not a safety harness and the vehicle safety belt must be worn over the top of the harness (Harness, 2017).

2.3   Transportation practices of CSHCN

A recent literature review on the transportation of CSHCN summarised the findings of 19 studies from the USA, Sweden, Australia and one study from Israel, however there were no studies conducted in Africa or from low and middle income countries (Downie et al., 2019). Thirteen of the studies were cross-sectional, three observational, one retrospective and one was a pre/post and follow-up design. Findings were summarized under the themes; CRS, wheelchairs, vehicles, travel habits, parental and professional knowledge and the study concluded that there is a strong need to increase knowledge of safe transportation of CSHCN as they continue to be inappropriately restrained (Downie et al., 2019). This review highlighted the need for observational research to provide a more accurate understanding on the transportation practices of CSHCN (Downie et al., 2019).

Studies investigating parental reported CRS use for CSHCN distributed questionnaires in different ways: mailed to parents (Falkmer & Gregersen, 2001; Falkmer & Gregersen, 2002), conducted telephonically (Huang et al., 2009; Huang et al., 2011), interview based with closed-ended questions (Korn et al., 2007), or conducted in conjunction with an observational study (O'Neil et al., 2009). Questions were grouped into themes and the following information was obtained by all investigators; CSHCN characteristics including age, gender and diagnosis, and transport details such as travel destinations, journey durations, vehicle type and choice of CRS (Falkmer & Gregersen, 2001; Falkmer & Gregersen, 2002; Huang et al., 2009; Huang et al., 2011; O'Neil et al., 2009). Other information that was obtained was specific to the research context or the research objectives.

For example two Swedish studies (Falkmer & Gregersen, 2001; Falkmer & Gregersen, 2002) investigated the self-reported knowledge of transport regulations and standards of families living in rural and urban areas, the CSHCN position in the vehicle, assistance required during transportation and whether additional passengers travel in the vehicle. Retrospective analysis of vehicle crashes in the USA (Huang et al., 2009; Huang et al., 2011) also considered the crash severity, direction of impact, injuries sustained by the CSHCN and drivers' particulars. Korn et al. (2007) conducted their research in Israel and included child ethnicity, highest level of education of the parent, behaviour and reported reason for CRS non-use. The survey conducted by O'Neil et al. (2009) focused more on comparing reported and observed practices of CSHCN transportation. They included more details about the seat position of the CSHCN within the vehicle, deactivation of airbags, modifications to CRS or vehicle and driver behaviour as a result of transporting a CSHCN. These questionnaires were not published or standardised therefore similar themes were used when developing a contextually relevant questionnaire for the study population.

Locally, an observational study investigated CRS usage, affordability and availability for children between 0 - 14 years old across seven suburbs in both hospital settings and childcare facilities in the Western Cape (Puvanachandra et al., 2020). Only 7.8% of children were observed using a CRS and 5.1% using a seatbelt, ensuing that 87.1% of children were travelling unrestrained (Puvanachandra et al., 2020). They highlighted the need for tighter seatbelt and CRS legislation and suggested the implementation of low cost/subsidised CRS or borrowing schemes and targeted social marketing to improve CRS usage (Puvanachandra et al., 2020). This study did not explore CRS misuse such as incorrect installation or early progression to the next type of device, nor did it take into consideration the specific needs and challenges faced by CSHCN in the South African context.

Previous research on sitting positions in CRS have investigated the child's posture during naturalistic driving, through the categorisation of head and trunk posture relative to the device (Andersson, Bohman & Osvalder, 2010; Charlton et al., 2010; Jakobsson et al., 2011). These studies provide an understanding of child behaviour, particularly relating to out of position (OOP) postures during transportation, for different age groups. However, to the best of the author's knowledge, there have been no studies that have investigated the seated posture of CSHCN in CRS during simulations or naturalistic driving.

Charlton et al. (2010) observed 25 children between 1 - 8 years old during naturalistic driving over a 3-week period. Children used their regular CRS, booster seat or harness in the vehicle, which had been checked by a CRS fitting specialist to ensure correct installation and fitment. Using video recording and analysis, they observed the direction that children moved OOP from the seats' centre line but did not measure the amount of deviation. In the same year, Andersson, Bohman and Osvalder (2010) investigated the effect of booster seat design on the child's choice of seating position. The study included of six children between the age of 3 - 6 years old and compared seated postures in two different CRS designs during car rides lasting 40 - 50 minutes. Their seating positions were observed through in-vehicle filming, categorised by OOP postural deviations from the midline and quantified the time spent in the deviated posture. Using the same categorisation as Andersson, Bohman and Osvalder (2010), but including an additional category in the sagittal plane for the slouched position, another Swedish study compared the sitting posture of six children between 8 - 13 years old in either a booster seat or with only a seatbelt during a 40 minute car drive (Jakobsson et al., 2011). In another study, the sitting postures, including slouching, were compared for six children aged 7 - 9 years old during an hour drive in two different CRS, the integrated booster cushion and the same Booster Seat used in Chapter 4: (Osvalder et al., 2013).

In order to quantify the range of head positions observed during naturalistic driving, Arbogast et al. (2016) incorporated the use of a Kinect™ sensor to analyse the head position of rear seat child occupants. The native Kinect skeletal tracking algorithm software suboptimally identified the head position and as a result the investigators developed their own algorithm (Arbogast et al., 2016). Their study analysed 135.5 hours of video obtained from 582 trips of 37 children aged 1 - 8 years old and found that as the CRS type moved from more to less restraint, the range of fore-aft and left-right head position increased (Arbogast et al., 2016). They proposed that the increased head movements were as a result of the booster seats and seatbelt only allowing more freedom of movement compared to a more restrictive car seat (Arbogast et al., 2016).

Together these studies included a broad range of children comprising of all the age categories recommended to use a CRS by the WHO except infants (World Health Organisation, 2009). Nevertheless, the study sample sizes were small resulting in possible bias due to higher variability (Simmons, 2018), and the results are susceptible to inflated effect size estimates (Button et al., 2013). The authors may have chosen smaller sample sizes due to the resource constraints of conducting a naturalistic driving study and categorising the video analysis (Arbogast et al., 2016; Arya, Antonisamy & Garg, 2012).

It is important to understand the relationship between child posture in a CRS and whether OOP postures result in increased injury risk (Arbogast et al., 2016). Using photo observations and virtual testing, van Rooij et al. (2005) investigated the injury potential of different child positions in a CRS for children between 1 - 3 years old. The posture study found that children tend to move around in their CRS during longer drives, resulting in slanted and slouched postures. Simulated load tests, using human child models representing 1.5 and 3 year olds, indicated that slouching and slanting resulted in increased neck loads (van Rooij et al., 2005). In one simulation, the child had escaped from the shoulder belt, and the head excursion far exceeded the 20cm limit, which in a vehicle would result in the child head impacting with the front row seat (van Rooij et al., 2005). Inappropriate belt fit may result in submarining, as the lap belt cuts into the abdomen while the pelvis slides underneath the belt, resulting in excessive abdominal penetration during crash tests (Brown et al., 2010). Obtaining data on occupants who assume positions in CRS that differ from the prescribed positions can lead to solutions for optimal protection (Arbogast et al., 2016), highlighting the need to observe seated postures in CRS in the CSHCN population.

2.3.1 Tools and instrumentation used to assess sitting balance and analyse seated posture in a car restraint system

Previous research was reviewed to find a sitting balance assessment, validated for CSHCN, that can be administered in a short period and require items commonly found in clinical practice to improve clinical utility. The assessment tool should also be able to be used with CHSCN with a range of intellectual impairments. The Sitting Assessment for Children with Neuromotor Dysfunction has been used in clinical practise to evaluate independent sitting ability in CSCHN diagnosed with CP (Knox, 2002). It is reliable for children between the ages of 2 – 10 who can sit without constant hand support however it requires videorecording each of the two five-minute phases of sitting (Knox, 2002), and would thus not be appropriate in this study.

A systematic review identified clinical tools to measure sitting posture, seated postural control or functional abilities in children with motor impairments (Field, D & Livingstone, 2013). The study identified 19 tools and reported on their reliability, validity and clinical utility but found that none met all the criteria for a well-developed outcome measure. Of these tools, only seven tools were identified for the assessment of sitting balance and posture, two of which focused on spinal alignment and were excluded for consideration in this study (Field, D & Livingstone, 2013).

The Segmental Assessment of Trunk Control, which requires custom strapping, and the Trunk Control Measurement Scale did not specify test duration but have more than 15 items usually indicating longer test durations (Field, D & Livingstone, 2013). The Box Sitting Ability portion of the Chailey Levels of Ability tool was found to be reliable and valid, however the 7-point ordinal scale does not describe the amount of support required by the CSHCN with poor balance (Field, D & Livingstone, 2013). The Trunk Impairment Scale, consisting of 17 items, but requiring only 10 minutes to administer requires the participant to understand instructions and sit on the bench without support (Pham et al., 2016). Lastly the Level of Sitting Scale (LSS) is an 8-point ordinal scale that can be completed in less than 10 minutes requiring only a bench (Field, D & Livingstone, 2013).

In the same year, Saether et al. (2013) also did a systematic review on clinical tools to assess balance in children and adults with CP. Of the 22 clinical tools that they reviewed only four focused on sitting: The Seated Posture Control Measurement, requiring an inclinometer and 20 – 40 minutes to complete the 22 items, the Sitting Assessment Scale which did not assess posture but rather compared posture between two sitting positions, the Level of Sitting Ability (LSA) and the LSS (Saether et al., 2013). The LSA is a 7-point ordinal scale that requires a short duration to complete and categorises sitting balance from unplaceable to independent sitting (Green & Nelham, 1991). The LSS is in fact adapted from the LSA, providing more detail on the amount of support required by children with difficulties with sitting balance (Saether et al., 2013). The LSS provides discrete measurable descriptions of sitting abilities, it is reliable, had its content validated and consists of an 8-point ordinal scale that requires only a bench and takes minutes to complete (Field, D & Livingstone, 2013). Hence the LSS was opted for use in this study to assess difficulties in sitting balance among CSHCN.

2.4   Conclusion

Poor postural support has been identified as a problem for CSHCN of all ages, and is a worry for many parents when transporting their CSHCN (Falkmer & Gregersen, 2002). In the event of a crash, CSHCN face an increased risk of injury and fatality, due to suboptimal restraint, compared with other children (Baker et al., 2012). A CRS provides protection to the child by restraining them within the vehicle and preventing or reducing dangerous contact with the vehicle interior (Brinkey, Manary & Santioni, 2011). Although some CSHCN can be safely accommodated in standard CRS (Baker et al., 2012; Falkmer & Gregersen, 2002), it might be unsafe or inappropriate for others (Baker et al., 2012), for which more specialised

CRS are needed. In South Africa, specialised CRS must be imported at considerably high prices as they are not locally produced.

Local research has attributed the low CRS usage amongst children in South Africa to the lack of accessible and affordable CRS (Puvanachandra et al., 2020), however no research has been conducted in the CSHCN population. Additionally, there are number of differences in the public transport systems between first world countries and developing countries such as South Africa (Walters, 2013). Therefore, research in developed countries cannot be generalised to low- and middle-income countries like South Africa, for which research investigating the need for specialised CRS within the South African context is warranted.

Chapter 3: Assessing the use of car restraint systems in children with special health care needs; a Western Cape based survey study

3.1 Introduction

Provision of transportation for CSHCN is complex, largely understudied (Korn et al., 2007) and faces many barriers including insufficient space to transfer in and out of the vehicle (Petzäll, 1995), long travel durations and time in the vehicle (Department of Basic Education, 2015). The majority of children with motor impairments require vehicular transport to travel to school, hospital or for recreational activities, which should be safe and comfortable (Baker et al., 2012). Children with physical disabilities might require special consideration when being transported due to their additional physical needs (Lindner, 2011). Children with poor postural control may need adaptive seating to improve postural support and sitting ability within the vehicle (Ryan & Rigby, 2007). Standard CRS which provide specialised protection, significantly reducing the risk of serious injury in an accident (Zaza et al., 2001) might therefore be unsafe or inappropriate for CSHCN.

In South Africa, a small number of specialised CRS are available for CSHCN. Although they provide multiple levels of postural support, customisation and ease of access, specialised CRS are imported, and most are considerably expensive (Sitwell Technologies, 2017). There are no locally produced CRS for CSHCN in South Africa. However, there are in-vehicle posture support systems which must be used in conjunction with the vehicles safety belt system as they have not been crash tested and thus not a safety seat (Shonaquip, 2017). Due to the high cost of specialised CRS and the need for additional postural support more than that provided by the standard CRS, South African parents may seek alternative seating solutions.

As no local research has been done on the transportation needs of CSHCN, it is important to assess the current CRS usage, and the challenges experienced by the CSHCN population within South Africa.

3.2 Aims and objectives

The study aimed to assess the current CRS usage as well as the parents' experiences and perspectives of transportation of CSHCN in the Western Cape, using a self-designed survey.

The specific objectives of the validation and pilot phase of the study were to:

i.   Evaluate and improve face validity of the questionnaire through focus group discussions,

ii.   Evaluate the feasibility of the recruitment process and survey procedure.

The specific objectives of the survey study were to determine, using a self-designed, validated questionnaire completed by the parents/caregiver of CSHCN:

i.   the modes of transport used by CSHCN,

ii.   the prevalence of the use of postural support systems during transportation for CSHCN,

iii. the current use of seatbelts, standard car restraints and/or adapted systems for the transportation of CSHCN,

iv.   the challenges faced by parents during CSHCN transportation.

3.3   Methodology

3.3.1   Study design

The descriptive quantitative survey study, including a pilot study, was performed at three public special schools offering daily transport services to their enrolled CSHCN, within the WCED Central district as described in the research setting (Chapter 1.4).

Prior to the main survey study, three focus group discussions were conducted, one at each of the included schools, to critically analyse, review and adjust the self-designed questionnaire, ensuring the questionnaire was valid (Appendix 7.1).

McManus et al. (2006:186) defined focus groups as being able to, "generate the subjective views of a group of individuals and allow exploration and reporting of all issues relevant to the subject". In their discussion groups with parents of CSHCN in Europe, they created cross-cutting themes that were relevant to their context. Focus group discussions comprising of a convenience sample have been used as a method of validity testing for self-designed questionnaires (Grant et al., 2007; Hayes, Fitzgerald & Jacober, 2008). Face validity determines whether the items of a self-designed outcome measure are sensible, appropriate and relevant to those using the measure (Connell et al., 2018).

3.3.2 Participants
3.3.2.1 *Focus group discussions*
Each school selected a convenience sample of six to eleven participants who were able to speak and understand English, including parents, bus drivers, class assistants and therapists. There was at least one representative from each group of participants. The participants were selected and invited by the school liaison based on their experience and knowledge of transporting CSHCN and availability during school hours. At the SID school there were eleven attendees: three bus drivers, three class assistants, two therapists and three parents. The CPSID centre had six attendees: one bus driver, one class assistant, one therapist and three parents. Lastly, the SLD school had nine attendees: three bus drivers, one class assistant, three therapists and two parents.

3.3.2.2 *Pilot survey study*
A sample population group was chosen by convenience and the parents of 65 learners from four classes in the Foundation Phase at the SID school, meeting the inclusion criteria of the main survey study, were invited to partake in the pilot study. The results of the pilot study are combined with the main study results as the minor changes made after the pilot phase did not impact the questionnaire significantly.

3.3.2.3 *Main survey study*
All parents/guardians of the remaining 600 learners enrolled at one of the participating schools, between the age of 4 – 18 years and able to read and understand either English or Afrikaans as per school language policies, were eligible for enrolment in the study. The parents/guardians who did not complete the transportation section of the questionnaire were excluded from the study. A sample of convenience was used.

3.3.3 Assessment tools
Questionnaires are an essential method to obtain information from a target population (Parfitt, 2005), and should be developed for the target context (Krosnick, 2018). Questionnaires can be easily delivered to a larger number of participants compared to an interview which also requires more time to conduct and transcribe (Adams & Cox, 2008). A self-designed questionnaire for parents/caregivers of learners was used to record demographic, financial and medical information, information regarding the transportation of learners, the parents' perspective of legislation and the use of CRS (Appendix 7.2).

The original version of the questionnaire, consisting of 19 questions, was created during the study's protocol development and validated for content by three experts in the field of paediatrics (Appendix 7.1). Content validity is the extent to which the items in a self-designed outcome measure comprehensively covers the different components to be measured (Connell et al., 2018). The original questionnaire had four mutually exclusive categories: 1) Understanding your child's disability 2) Transportation 3) Legislation 4) Social Circumstances. The questionnaire was thereafter validated through focus group discussions with stakeholders from each participating school for face validity. Amendments were made based on the suggestions from the focus groups (Chapter 3.4.1.1 and Appendix 7.1). Resulting in a 28 item, validated, self-designed questionnaire used in the pilot phase and main study (Appendix 7.2). A COSMIN checklist can also be found in Appendix 7.3 to describe the measurement properties that were assessed.

3.3.4 Study procedure

A diagrammatic representation of the survey study's procedure can be found in Figure 5. Ethical approval was granted by the Faculty of Health Sciences Human Research Ethics Committee, University of Cape Town (Appendix 7.4). Furthermore, permission to conduct research at the selected schools was obtained from the WCED (Appendix 7.5).

Principals and/or governing bodies of participating schools were approached (Appendix 7.6) to obtain oral permission to conduct research at their institution, to provide access to their learners and to allow school therapists to assist in gathering information.

3.3.4.1 *Focus group discussions*

A convenient date for the focus group discussion was set with the school liaison and each school provided a private venue with tables and chairs set up in a boardroom style to facilitate discussion amongst the participants. Informed consent to partake in the study was obtained from all participants prior to commencement of the focus group (Appendix 7.7). Each participant received a copy of the questionnaire and a pen to make additional notes if required.

Focus groups lasted between 60-90 minutes and were recorded electronically via a voice recorder to easily transcribe the agreed amendments. After participants were welcomed, the aims of the research study were explained and linked to the purpose of the focus group discussion. Time was given for each participant to read three to five questions at a time, after which each question and the proposed answers were discussed. They were encouraged to try and answer the questions keeping in mind a CSHCN that they knew to ensure context and critical thinking.

The groups discussed the mechanics and semantics of each question, their understanding of terminology, the appropriateness of the question and the likelihood of the answers to adequately describe the problem. Suggestions were given to amend the questionnaire and it was updated after each focus group discussion. At the end of each focus group discussion, there was consensus amongst the participants that the revised questionnaire covered all appropriate aspects of the transportation needs of the learners.

3.3.4.2 *Pilot study*

The pilot study, conducted over a six-week period, followed the procedure of the main study explained below to determine the feasibility of the recruitment process and implementation of the survey study.

3.3.4.3 *Main study*

The informed consent form (Appendix 7.8) and the self-designed parent questionnaire on learner transport (Appendix 7.2) were distributed to parents of eligible learners at the schools via the learners' message books, the preferred communication method at the schools. The schools' language policies requested that documentation be distributed in English or Afrikaans, hence translation to isiXhosa was not found necessary. The questionnaire was translated by an independent translator and proof read for any adjustments by a second translator. Teachers were responsible for following up with parents to ensure consent forms and questionnaires were returned within six weeks.

3.3.5 Data capturing and management

Returned questionnaires were collated per class by the class teacher, handed to the investigator in sealed envelope to ensure confidentiality after which the questionnaires were coded and de-identified. Data were entered into a password protected Excel document and imported to statistical software, Statistica 13 (TIBCO Software Inc, 2018), for further analysis. All hard copies were locked in a cupboard and will be destroyed after book submission. Electronic data will be kept for a minimum of two years as per Health Professions Council of South Africa Ethics Booklet 13 (Health Professions Council of South Africa, 2008).

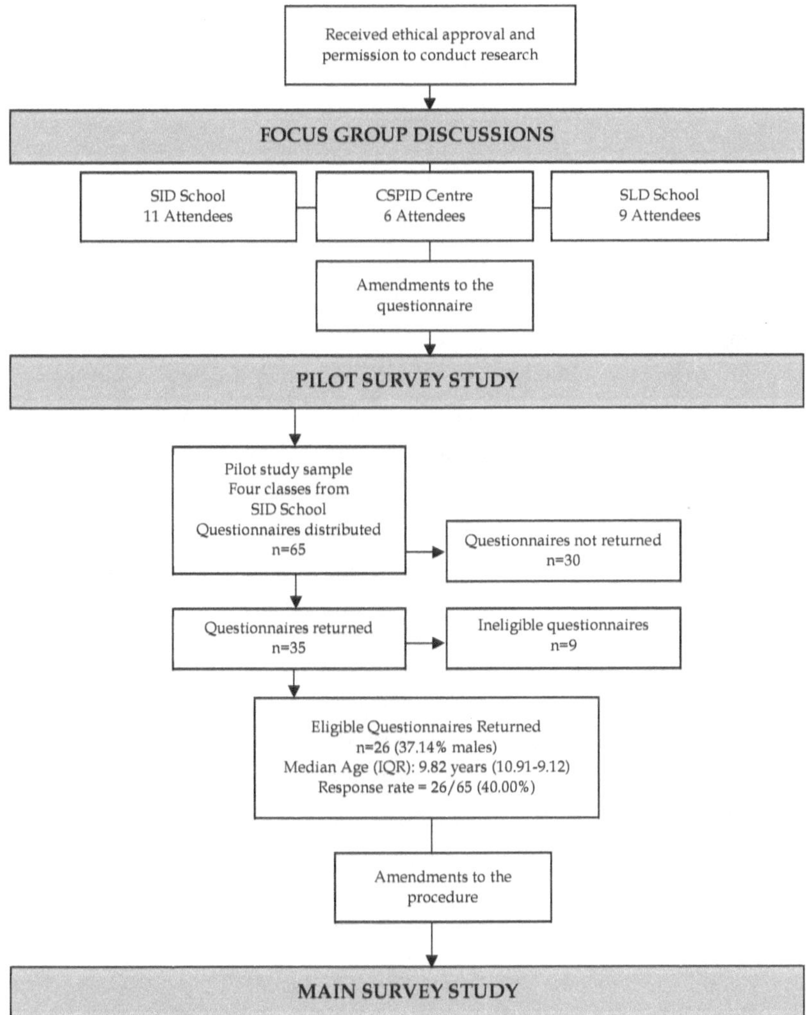

Figure 5: Flowchart showing procedure for the survey study and sample populations for the pilot survey study

3.3.6   Data analysis

Nominal and ordinal data such as age, weight and height were tested for normality with the

Shapiro-Wilk test (Ghasemi & Zahediasl, 2012), for which data were presented as median

and interquartile range (IQR). An analysis of age compared to CRS usage was done with the

Mann-Whitney U test. The frequency distribution was examined for both nominal and ordinal data and presented as n-value and percentages. Associations between categorical variables in the questionnaire were determined using Chi-square with Fisher's exact or Yates correction in case of small sample sizes. A significant association was determined by a p-value of <0.05.

3.4 Amendments

3.4.1.1 *Amendments as a result of focus group discussions*

The original 19-item questionnaire was reworked into 28 questions. Amendments to the questionnaire post-focus group discussions included correcting numbering and formatting errors, adding simple explanations for medical terminology as well as improving the structure and order of the questions for improved flow of information. To improve the quality of the answers provided and ensure ease of understanding, some of the questions were expanded or divided into multiple questions. The option of "none" or an opportunity to describe the problem was given. Further descriptors on challenges faced by the parents taking into consideration the children with more severe disabilities were considered. Details of the adjustments made as a result of each focus group can be found in Appendix 7.1.

Despite common amendments to the mechanics and semantics of the questionnaire, distinct themes and concerns with regards to the content were identified from reviewing the focus groups at each of the schools. The SID school with a wide range of severity in their learners' disability, emphasised elaborating on the questions to provide an opportunity to better describe the circumstances experienced by parents during their daily routine. The group from the CPSID centre concentrated on accommodating the challenges faced in transporting children with severe disabilities. The priority for the group from the SLD school, where the learners had mild intellectual disabilities and were socially more aware, was focused on how the challenges were perceived by the child within their context.

The diverse feedback from these focus group discussions reaffirms the decision to select schools from the different types of special schools within the WCED. The feedback facilitated the amendments made to the parent questionnaire, ensured all learners were represented in the questions and answer options, and ensured face validity of this self-designed tool.

Minor adjustments were made to the format of the informed consent form to ensure clearer interpretation when completing and signing the form. It was recommended that consent forms and questionnaires be stapled together when sent to parents to encourage a higher return rate of both forms and questionnaires.

After concluding the pilot survey study, it was decided to include an additional category between low and middle income when enquiring about total monthly household income to prevent category bias.

A trend was noted that parents who experienced few challenges came across a question pertaining to the severely disabled child and skipped that question as well as the rest of the page. As such the order of some of the questions were adjusted to prevent omission within the questionnaire and to facilitate more comprehensive responses from parents. It was also decided to include an additional category for total monthly household income to prevent category bias.

3.5   Results

The demographic characteristics of the pilot survey study were analysed separately and in conjunction with the main survey study to illustrate the sample populations. All other results are presented in combination.

3.5.1   Demographic characteristics

A total of 665 questionnaires were distributed (65 for the pilot study and an additional 600 thereafter). Only 277 questionnaires were returned (35 in the pilot phase), of which nine were ineligible. Therefore the final sample included 268 questionnaires (40.30% eligible return rate) (Figure 6). The response rates were evenly distributed among the schools (Figure 6), however, because the number of learners enrolled in each institution varies, the respondents from the SLD school (n=128) represent 47.76% of this sample. The age of the participants skewed towards older CSHCN in the pilot phase (W=0.778 ; p<0.001) and the main study (W=0.971 ; p<0.001). The median (IQR) age of the learners was 11.52 (14.63-8.86) years in the main study and 9.82 (10.91-9.12) years in the pilot phase (Figure 6). There were 37.14% (n=10) male respondents in the pilot phase and 58.96% (n=158) in the main study (Figure 6).

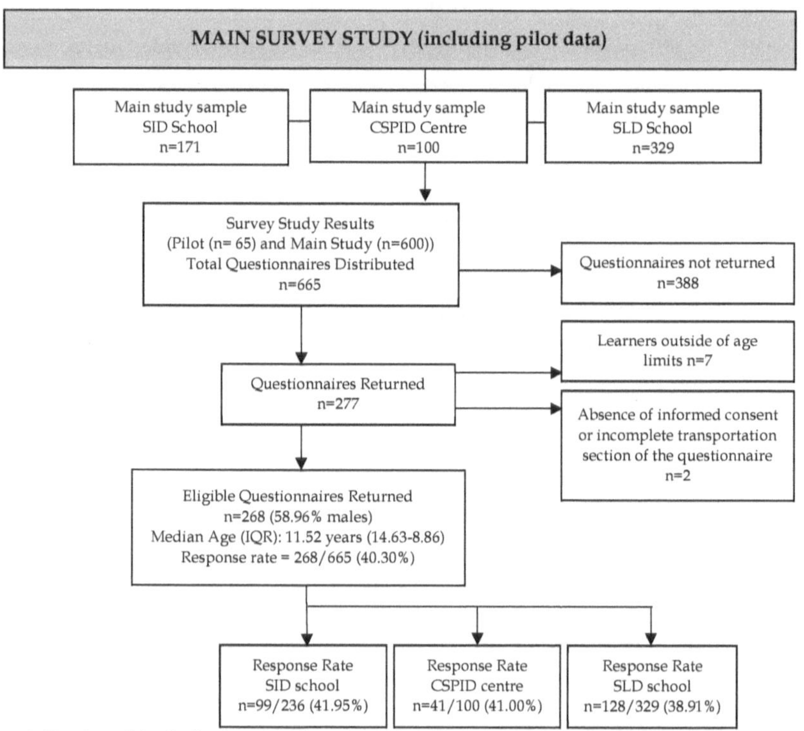

Figure 6: Flowchart of the distribution of questionnaires included in survey study as well as presenting response rate, in percentage, and demographics

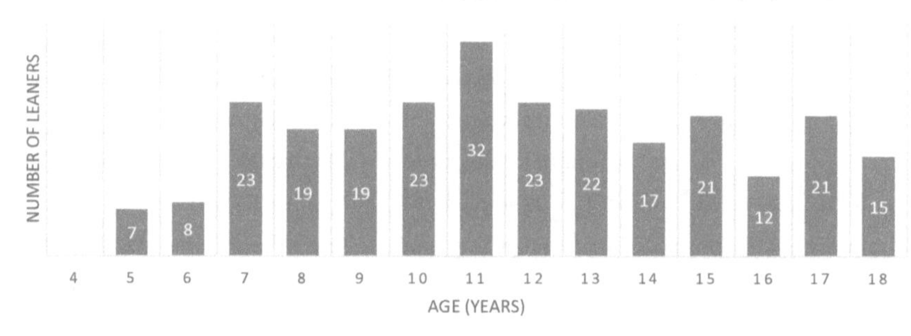

Figure 7: Age distribution of participants (years), based on n=262

Participant diagnoses included neurological conditions, genetic syndromes and motor dysfunction (Table 2); the most common being cerebral palsy (n=78, 29.10%). The highest reported impairment impacting the use of CRS was spasticity (n=77, 28.73%) (Table 2).

Table 2: Reported diagnosis and secondary complications impacting the use of child restraint systems in children with physical disabilities, based on n=268.

Diagnosis/Secondary Complication		n (%)
Diagnosis	Cerebral Palsy	78 (29.10)
	Genetic Syndrome	10 (3.73)
	Spina Bifida	8 (2.99)
	Mild Motor Problem	7 (2.61)
	Upper Motor Neuron Lesion	5 (1.87)
	Muscular Dystrophy	4 (1.49)
	Other	9 (3.36)
	Unknown	147 (54.85)
Impairments impacting the use of CRS	Spasticity	77 (28.73)
	Scoliosis	23 (8.58)
	Joint Contracture	15 (5.60)
	Hip Dysplasia	9 (3.36)
	Loss of Sensation	8 (2.99)

3.5.2 Modes of transportation and other factors related to the transportation of children with special health care needs

Of the 268 participants, 73.13% (n=196) were transported to school using public transport (Figure 8). The types of public transport used were school bus (55.20%, n=148), school minibus (13.80%, n=37) and others including taxi and train (4.1%, n=11). However, when travelling outside of school hours for short distances, families opted to either walk (38.43%, n=103) or use private transport (31.72%, n=85) (Figure 8). When needing to travel further distances 55.60% (n=149) made use of private transport followed by 26.87% (n=72) using

public transport (

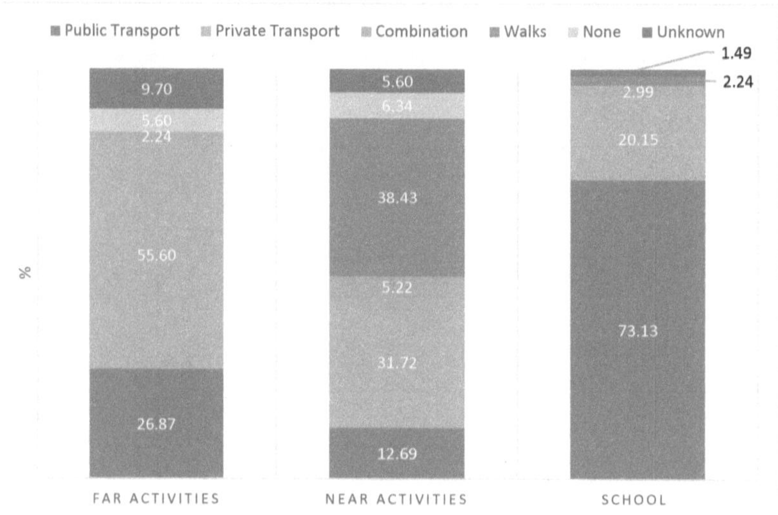

Figure 8). Nearly half of the participants (44.03%, n=118) reported spending more than an hour in a vehicle daily. Thirty eight percent (n=102) of parents felt their child required supervision during the journey, either to support an upright posture for the child or to manage their behaviour.

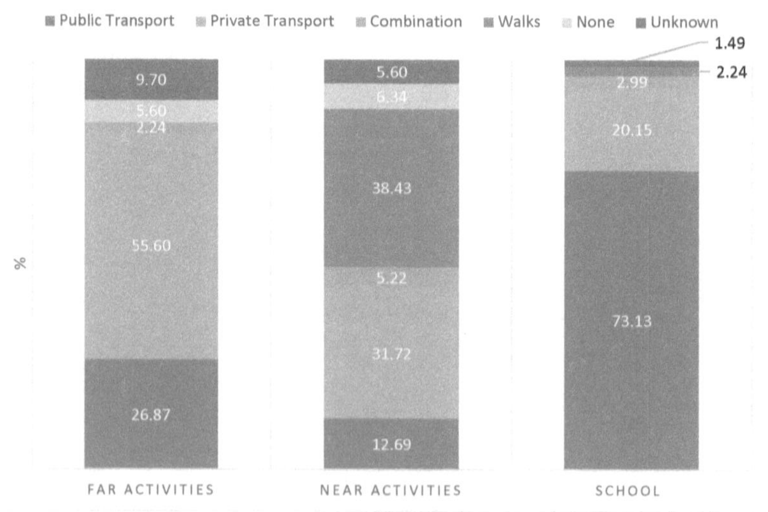

3.5.3 Prevalence of the use of postural support systems during transportation

While 66.04% (n=177) of participants use a seatbelt during their commute, 6.34% (n=17) reported misuse, by adapting the CRS, or that CSHCN are unrestrained and 14.18% (n=36) did not report on seatbelt or CRS use (Figure 9). The remaining CSHCN either use a standard CRS (n=28, 10.45%), specialised CRS (n=4, 1.49%) or remain in their wheelchair in an adapted vehicle (n=4, 1.49%). Comparison of the age of children using and not using a CRS revealed that CRS was used by younger CSHCN (Mann-Whitney U=1830.50, p=0.004) (Figure 10).

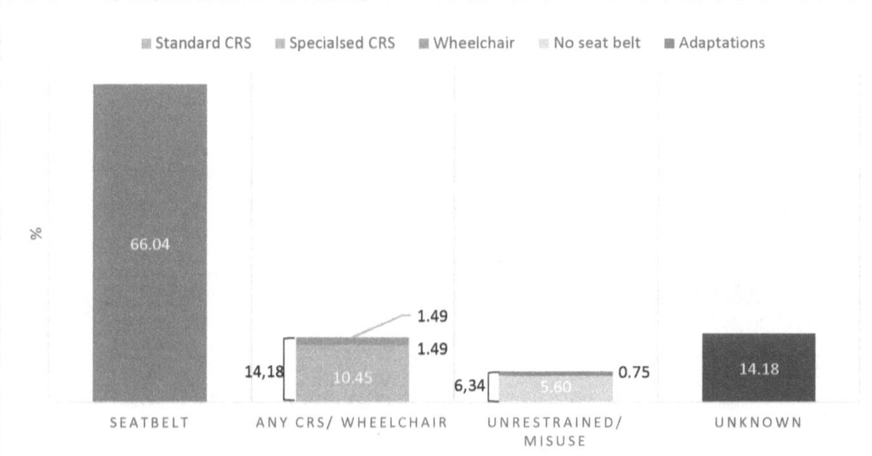

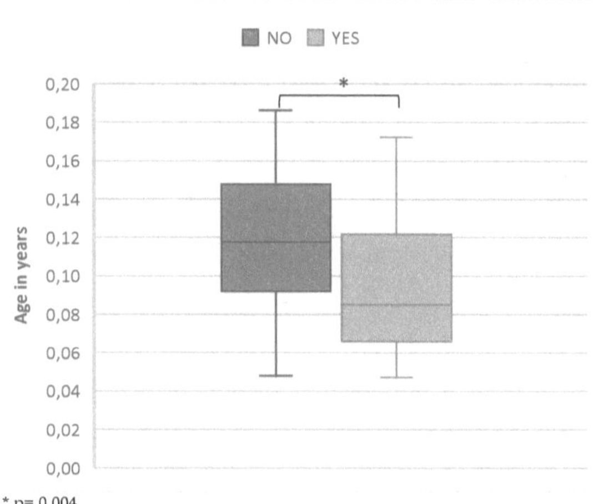

* p= 0.004

Figure 10: Boxplot comparing the age of children using and not using CRS, based on n=268.

Sixty nine parents (25.75%) reported difficulties with their child's sitting balance, ranging from limited movement and balance in sitting to the need for additional postural support

Table 3). Regardless of their sitting balance, all children in their early years should use CRS,

Parent perception of sitting balance		n (%)
Independent sitting ability	Any type of chair	171 (63.81)
Difficulties with sitting balance	Independent static sitting balance only	24 (8.96)
	Requires backrest or uses hands for support	13 (4.85)
	Requires well-supported chair with backrest and armrests	22 (8.21)
	Requires maximum support in sitting	10 (3.73)
	Subtotal: Difficulties with sitting balance	69 (25.75
Unknown	No response	28 (10.45)

however, only 41.79% (n=112) of parents have ever appropriately made use of a CRS for their child (Figure 11). The parent's perspective of their child's difficulties with sitting balance was significantly associated with the use of CRS (X^2= 17.72, p< 0.001) (Table 4), as parents who felt that their child had difficulty with their sitting balance were more likely to use a CRS. Furthermore, a significant association was observed between currently using a CRS and child's weight (X^2= 11.54, p=0.021) (Table 4), as children who weighed more were less likely to still be using a CRS. Anthropometric data showed that approximately half of the children were still within the recommended weight and height criteria for standard CRS (55.22% (n=148) and 48.88% (n= 131), respectively), compared to 8.96% (n=24) of children that are still making use of a CRS (Figure 11).

Table 3: Parents' perception of child's sitting balance abilities (in %), based on n=268

Parent perception of sitting balance		n (%)
Independent sitting ability	Any type of chair	171 (63.81)
Difficulties with sitting balance	Independent static sitting balance only	24 (8.96)
	Requires backrest or uses hands for support	13 (4.85)
	Requires well-supported chair with backrest and armrests	22 (8.21)
	Requires maximum support in sitting	10 (3.73)
	Subtotal: Difficulties with sitting balance	69 (25.75
Unknown	No response	28 (10.45)

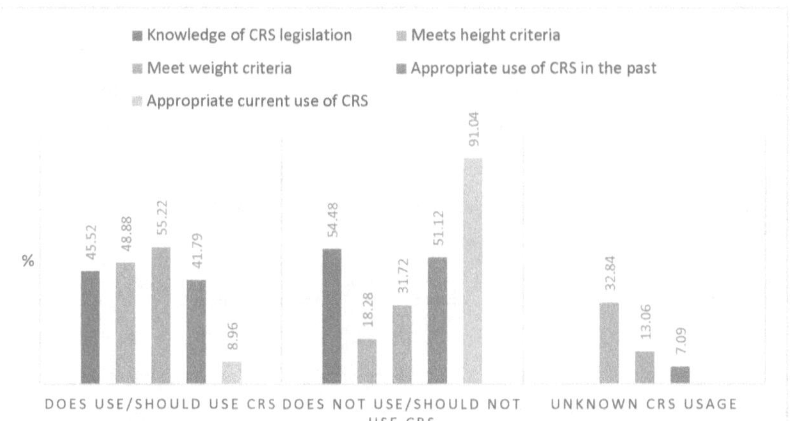

Figure 11: Comparison of parents' knowledge of car restraint legislation, anthropometric requirements, the appropriate use of CRS currently and historically (in %), categorised by CRS usage, based on n=268

Table 4: Observed frequencies and two-way summary table for parent perspective of child's difficulty with sitting balance (n=240) and child's weight to use of CRS (n=233), Yates corrected

Category	Subcategory	Does not use a CRS	Still uses a CRS	Row TOTAL	Statistical & p-value
Parent perspective of child's difficulty	Any type of chair	161	10	171	X^2= 17.72
	Independent sitting	19	3	22	p< 0.001
	Backrest or support	21	3	24	based on
	Well-supported chair	5	5	10	n=240

with sitting	Maximum support	10	3	**13**	
ability	**Totals**	**216**	**24**	**240**	
Child's	<13kg	1	1	**2**	X²= 11.54
weight	13kg - 18kg	25	8	**33**	p= 0.021
	18kg - 36kg	98	13	**111**	based on
	<36kg	2	0	**2**	n=233
	>36kg	83	2	**85**	
	Totals	**209**	**24**	**233**	

3.5.4 Factors influencing car restraint system use amongst children with special health care needs

The majority of children (n=158, 58.96%) came from low-to-middle income households (<R6500 per month per household) (Table 5). The mean (SD) number of persons supported by a low-to-middle monthly household income of R2 500 – R6 500, was 7.09 (4.32) (Table 5), reducing their per capita income. Household income significantly impacts the use of a CRS, as most children from higher income families are transported in a CRS (X²= 48.14, p< 0.001) (Table 6).

Half of the parents (n= 139, 51.87%) were not willing to spend money on a CRS as they felt that a car seat was not necessary for their child. Only 5.97% (n=16) were willing to spend sufficiently for a standard car seat, while none were able to consider spending sufficiently towards the current cost of a specialised device for CSHCN (Figure 12). The amount that parents were willing to spend towards a CRS was significantly associated with having ever made use of a CRS (X²=43.38, p<0.001) (Table 6).

The current South African legislation on vehicle safety restraints is not known by all parents. There were 18.66% (n=50) who did not know the legislation on seatbelts and 54.48% (n=146) who did not know the CRS law (Figure 11). Knowledge about CRS legislation has a significant positive association with the use of a CRS (X²= 19.84, p<0.001) (Table 6).

Table 5: Participant monthly household income categorised by income brackets (in %) and the mean number of persons supported by this income, based on n=268

Monthly household income		n (%)	Mean (SD)
Low-to-middle income	<R 1600	34 (12.69)	4.25 (2.22)
	R1 600 – R2 500	46 (17.16)	5.75 (2.95)
	R2 500 – R6 500	78 (29.10)	7.09 (4.32)
	Subtotal: <R6 500	158 (58.95)	5.85 (3.36)
Middle income	R6 500 – R15 600	36 (13.43)	4.00 (2.11)
High income	>R15 600	41 (15.30)	4.56 (4.06)

Unknown	33 (12.31)	1.00 (0.00)

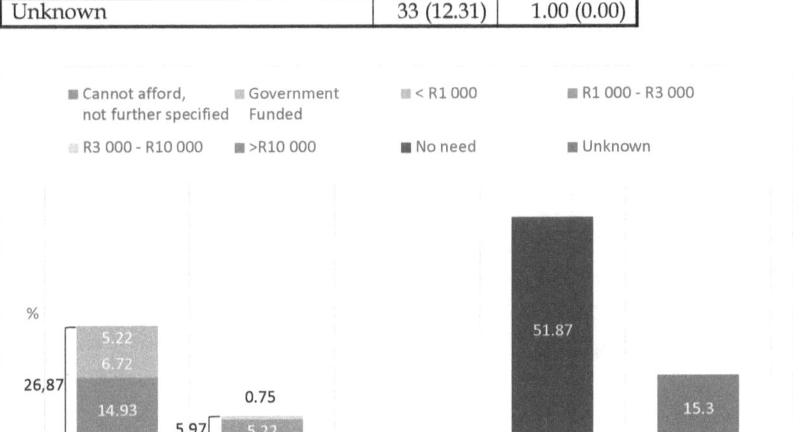

Figure 12: Parent's perception about the need for and availability of funds for a CRS (in %), based on n=268

Table 6: Observed frequencies and two-way summary table for household income (n=221), amount willing to spend on a CRS (n=133) and knowledge of CRS legislation (n=226), Yates corrected

Category	Subcategory	Does not use a CRS	Still uses a CRS	Row TOTAL	Statistical & p-value
Household income	<R1600	22	7	29	X^2= 48.14 p< 0.001) based on n=221
	R1600 - R2500	33	9	42	
	R2500 - R6500	49	27	76	
	R6500 - R15600	17	19	36	
	>R15600	3	35	38	
	Totals	124	97	221	
Amount willing to spend on a CRS	<R1000	2	12	14	X^2= 43.38 p< 0.001 based on n=217
	R1000 - R3000	0	14	14	
	R3000 - R10000	1	1	2	
	None – cannot afford	26	11	37	
	None – government should subsidise	2	15	17	
	My child does not need a CRS	88	45	133	
	Totals	119	98	217	
Knowledge of CRS legislation	Inadequate	74	33	107	X^2= 19.84 p< 0.001 based on n=226
	Accurate	46	73	119	
	Totals	120	106	226	

3.5.5   Challenges faced by parents during transportation

The mode of transport, each coming with their own challenges, was linked to the distance travelled and affordability (see Table 7). Parents using private transport indicated that transporting the wheelchair (n=29, 10.82%) and the unavailability of demarcated disability parking bays (n=20, 7.46%) are the main transportation challenges. Amongst parents using public transport, the greatest challenges were the child's poor sitting balance (n=17, 6.34%) and lack of space (n=15, 5.60%) within the vehicle. The main reasons preventing parents to travel with their CSHCN, regardless of the mode of transport, was the affordability of transport services (n=22, 8.21%) and the inability to accommodate their child's disability (n=11, 4.10%) during transportation.

Table 7: Challenges experienced by parents when transporting children with physical disabilities by mode of transport, based on n=268

Challenges faced by parents of CSHCN by mode of transport		n (%)
Private transport	Wheelchair too bulky in the vehicle	29(10.82)
	Lack of available/demarcated disabled parking bays	20(7.46)
	Child's weight	15(5.60)
	Lack of space within the vehicle due to seat close together	13(4.85)
	Lack of space around the vehicle to park wheelchair nearby	12(4.48)
	Difficult to load child in the vehicle due to the size of door	11(4.10)
	Child's ability to sit independently	9(3.36)
Public transport	Child's ability to sit independently	17(6.34)
	Lack of space in the vehicle for child and their wheelchair	15(5.60)
	Insufficient time to load into the vehicle	11(4.10)
	Wheelchair too bulky in the vehicle	11(4.10)
	Difficult to load child in the vehicle due to the size of door/pavement height	10(3.73)
	Child's weight	9(3.36)
	Lack of space within the vehicle due to seat close together	8(2.99)
	Driver refuses to transport child with disabilities/wheelchair	4(1.49)
Challenges resulting in CSHCN not travelling	Affordability	22(8.21)
	Child's needs, including wheelchair, cannot be accommodated	11(4.10)
	Wheelchair too bulky in the vehicle	10(3.73)
	Access to public transport	7(2.61)

	It is unsafe	5(1.87)
	It takes too long to prepare the child for travelling	4(1.49)
	Difficult to load the child in the vehicle	4(1.49)

3.6 Discussion

To the best of the authors knowledge, this is the first study in an African country which determined the prevalence of postural support systems usage during transportation of CSHCN and explored the challenges faced by their parents. This study had an overall response rate of 40.30%, resulting in possible non-response bias, although it falls between the varied response rates of similar studies conducted in the USA (27%), Sweden (81%) and South Africa (58.3%) (Everly et al., 1993; Falkmer & Gregersen, 2001; Puvanachandra et al., 2020). It is also similar to other South African studies which distributed questionnaires amongst teachers (56%) (Mathews et al., 2006) and doctors (29.5%) (De Vries & Reid, 2003). Randomised control studies suggest that telephone prompt before questionnaire distribution and cash incentives may improve response rate (Locker, 2000; Whiteman et al., 2003). Additional strategies could be implemented in future research to improve response rates and reduce the potential non-response bias.

3.6.1 Sitting balance and the postural support requirements of children with special health care needs

The most common diagnosis amongst study participants was CP (29.10%), as anticipated by the schools' enrolment policies. This is similar to a questionnaire based transportation study in Sweden (56%) (Falkmer & Gregersen, 2001), and an observational study conducted in Israel (37%) (Korn et al., 2007), where CP was also the most prominent diagnoses. The most common physical disability in children is CP (Rosenbaum, 2003), supporting the enrolment of these CSHCN in public special schools.

CSHCN, particularly those with CP, may present with postural dysfunction and find it challenging to support their own torso or head in the upright seated posture (Carlberg & Hadders-Algra, 2005; Lindner, 2011). This can result in a poor sitting posture or difficulty achieving balanced sitting postures (Park et al., 2001). Twenty-six percent of parents in the current study reported difficulties with their child's sitting balance, ranging from limited independent movement while sitting, to the need for additional postural support due to poor sitting balance. However, the child's sitting balance has not been the focus of other research on the transportation CSHCN (Falkmer & Gregersen, 2001; Korn et al., 2007). O'Neil et al. (2009) observed that CSHCN with poor head and trunk control may require additional postural support in the vehicle from CRS and/or additional padding. Furthermore, associated physical conditions such as hypotonia or scoliosis may also affect their ability to fit in a CRS (Huang et al., 2011).

3.6.2 Modes of transportation

Our results showed that two thirds (69.00%) of the participating CSHCN relied on school transport by buses or mini-buses for their daily commute to school. This is higher than international surveys on the transportation of CSHCN to school (43.6 – 46%) (Falkmer & Gregersen, 2001; Wheeler, Yang & Xiang, 2009). A survey conducted in the United States of America (USA) found that a bus was the most common mode of transportation to school (43,6%) but did not differentiate between private, public or school owned vehicles (Wheeler, Yang & Xiang, 2009). Similarly, Falkmer and Gregersen (2001) found that most CSHCN (46%) in their survey used school transport, but they too did not describe the type of vehicles used. A recent systematic review by Downie et al. (2019) found that the large majority of CSHCN travelled in school transport or private vehicles. As most CSHCN in this study utilise school transport it is important that these vehicles are equipped with the necessary CRS to ensure safety and comfort.

When travelling outside of school hours, the current study found that for shorter distances, walking (38.43%) and using private transport (31.72%) had similar frequencies; and private transport (55.60%) was used twice as often as public transport (26.87%) in longer distances. Regardless of the reason for personal travel, surveys conducted in the USA and Sweden showed that private transportation was the most common mode of transport (between 58 - 96%) (Falkmer & Gregersen, 2001; Wheeler, Yang & Xiang, 2009). In both studies, more than 95% of families owned a private vehicle. Access to a private vehicle in South Africa is associated with higher household incomes (Walters, 2013), suggesting that families with a low monthly household income would need to seek alternative transport. Although this study did not investigate vehicle ownership, only 29% of South African households owned a car in 2013 (Jeske, 2016), indicating that the majority of families depend on the public transport system.

In terms of accessibility, safety and CRS usage, public and private transport each have benefits and challenges. An advantage of private vehicles is that there are seatbelts in every seat for the installation of CRS however only 55.60% of families in the current study use this mode of transport when travelling far distances and 31.72% when travelling short distances. Public transportation, which is more commonly used in South Africa, is a considerably cheaper alternative (Walters, 2013) but regulations do not require fitment of a three-point seatbelt for CRS or specialised postural support systems installation (National Road Traffic Act, 1996). Public transportation in South Africa is often overcrowded (Walters, 2013), resulting in less space for bulky CRS or wheelchairs.

Whilst most respondents (44.03%) travel for more than an hour each day, 87% of Swedish parents reported less than an hour travel per day (Falkmer & Gregersen, 2001). A local investigation found that travelling with public transportation can take two to three times longer than trips in a private vehicle due to the added time of walking and waiting times to access public transport (Hitge & Vanderschuren, 2015). CSHCN face these same challenges of accessibility and long travel durations when using transportation services. The average travel time (90 minutes) in Cape Town is at the upper end of the global average (70 minutes) and is significantly longer for public transport users (110 minutes) due to a discrepancy in infrastructure (Hitge & Vanderschuren, 2015). Long travel durations could result in a higher exposure to accidents and a greater demand for comfortable seating solutions (Falkmer & Gregersen, 2001).

3.6.3   Restraint and postural support system use during transportation

This study sought to understand the frequency and type of restraint or postural support system, including a seatbelt used by CSHCN during transportation. Total restraint or postural support system usage in this study was 79.47%, with only 6.34% of parents reporting their CSHCN using no restraints at all and 14.18% did not disclose the restraint used. A similar survey-based Swedish study found that restraints were used by 99% of CSHCN (Falkmer & Gregersen, 2001). This may be due to the availability of several specialised CRS as well as a high degree of compliance to Swedish national road regulations (Falkmer & Gregersen, 2001).

Locally, and contrary to the current study's results, a paediatric trauma centre reports that 87% of child passengers involved in vehicle accidents were not adequately restrained and single-event observational studies found that between 85.1 - 89% were unrestrained (Kling, 2011; Puvanachandra et al., 2020), suggesting that restraint use is much lower than results reported in this study. This raises concerns about response bias as a result of overreporting.

Adults in countries with low belt use have also been observed exaggerating seatbelt use (Özkan et al., 2012; Parada et al., 2001).

A study by Korn et al. (2007) was the first to investigate observed versus parental reported CRS use in a CSHCN population. They advised that there are limitations to using parental reporting as an indicator of CRS use due to overreporting, particularly in populations of low CRS use. Their study showed 44.2% overreporting by parents and suggested that parents have a tendency to give a socially desirable response and one should be cautious to interpret information acquired from parents (Korn et al., 2007). Considering the inaccuracy of over-reporting by parents, it is concerning how many CSHCN are not safely restrained when travelling in vehicles, as correctly restraining a child significantly reduces risk of serious injury or death (Kling, 2011). It is suggested that CSHCN such as medical, orthopaedic, neuromuscular and behavioural needs may have a higher proportion of misuse or non-use of CRS (Korn et al., 2007). Although the current study only has parent reported CRS usage whereas other studies report on observed CRS usage, if overreporting is considered then one would expect the observed results of participants from this study to be lower for both CRS and seatbelt use.

When considering CRS usage alone, this study reported that only 13.43% of CSHCN used a CRS during transportation, whereas Falkmer and Gregersen (2001) reported 54% of CSHCN used a CRS in a private vehicle. CRS usage in this study may be lower due to the lack of availability of CRS on public transportation systems. Moreover, as these studies included CHSCN up to 16 or 18 years old, some CSHCN may have already exceeded the anthropometric or legislative requirements (World Health Organisation, 2018a) for a CRS and progressed to using a seatbelt only. Parents in this latter survey indicated the type of restraint in different modes of transport, however this was not investigated in the current study. Due to seatbelt configurations in different types of vehicles, restraint accessibility and subsequent use may differ between modes of transport. Studies in the USA have found that between 60.8 - 82% of CSHCN were appropriately restrained (Huang et al., 2011; O'Neil et al., 2009), indicating higher CRS usage rates compared to this study. CSHCN in South Africa may not be being transported adequately to adhere to recommended safety standards.

Our study included CSHCN up to the age of 18 years, as some CSHCN may still be within the height and weight limits of the CRS. It is anticipated that some of the older study participants would already exceed the anthropometric requirements of a CRS and would only require a seatbelt to be safely restrained. For this reason, we enquired about historical CRS use and only 41.79% of parents reported having ever used a CRS for their CSHCN. This suggests that the majority of South African CSHCN have never used a CRS, even as an infant, as is required by law (Department of Transport, 2014). It is therefore necessary to investigate the reason for this non-compliance amongst parents of CSHCN.

The current data showed that 55.22% of the CSHCN should still meet the anthropometric requirements of standard CRS but there are only 8.96% of participants that are currently using a standard or specialised CRS. This means that CRS misuse occurs in the 46.26% of study participants who are within the requirements for CRS but are not using one. It is unclear whether the misuse is due to lack of CRS availability, suitability or progressing to a seatbelt too soon. Therefore, further research may be required to determine the type of CRS misuse, the cause and how to prevent it in this population. An observational study on CSHCN found that all participants who were restrained in the vehicle displayed misuse, either in choice or use of the CRS (Korn et al., 2007). However, they did not quantify incorrect CRS choice alone compared to improper harness use. Premature graduation of CSHCN from CRS to adult seatbelts is common and if the incorrect restraint is used it can increase the risk of injury or death (Teerds & Cameron, 2015). It is recommended to use the child's size, developmental/behavioural characteristics and medical condition when determining the appropriate CRS instead of just age (Bull, 2008). Other considerations include the family financial circumstances, the number of other children being transported and the choice of family vehicle (Bull, 2008). An observational study in the USA reported that CRS misuse occurred frequently including CSHCN being progressed to the next type of CRS prematurely according to their weight or height (O'Neil et al., 2009). Every child should use a CRS until they exceed the manufacturers' height and weight recommendations to increase survival and reduce the severity of injury during a crash (World Health Organisation, 2009).

Two CHSCN (0.75%) in the current study report modifying the CRS to improve comfort, which is comparatively lower than an observational study showed that modifications to the CRS were reported for 24.1% of their CSHCN to improve the fit (O'Neil et al., 2009). A likely explanation for less modifications in this study is that the total number of CRS users was only 8.96%, resulting in less chance for parents to need to make a modification. Modifications to CRS can include placing padding under the cover, harness or behind the child as well as frame alterations to improve the support and comfort of the child but may alter the restraint protection during a crash (O'Neil et al., 2009). The latest guidelines published by the AAP recommend that CRS should not be modified in any manner unless specified by the manufacturer (O'Neil & Hoffman, 2019). It is mandatory for children to be adequately restrained and secured in the vehicle regardless of the destination, travel duration or disability (World Health Organisation, 2018b). however, the extent to which the CRS were modified in this study is unclear

3.6.4   Car restraint system use amongst CSHCN and the influencing factors

The study investigated the factors that influenced the use of CRS and found that higher household income, the child's weight, parents' perspective of their child's difficulty with sitting balance and knowledge of current legislation are important factors of influence. Which has some similarities to a recent review, which found that challenges to CRS use in CSHCN include the cost of purchasing a CRS, unavailability of adequate CRS that accommodate the needs of the CSHCN, vehicle overcrowding preventing the use of CRS, the child's resistance to being restrained and lack of parental awareness (Downie et al., 2019).

While most responding parents (51.87%) reported that they had no need to purchase a CRS, 26.87% were unable to afford the cost of a standard CRS and no family was willing to spend the amount of money necessary for the high cost of a specialised CRS. This may be due to the low monthly household income (58.95% earning less than R6 500/month). CSHCN requiring specialised CRS might not therefore be able to afford these devices without additional financial support. Funding options can include charities, non-government organisations (Baker et al., 2012), or government funded strategies. South African families do not have the financial resources due to lower average monthly household income resulting in greater dependence on external funding from private or government stakeholders.

A survey on families with CSHCN in Australia found that 74.7% were able to self-fund the recommended CRS, the others continue transporting the child in a manner that is considered unsafe (Baker et al., 2012). However, Wheeler, Yang and Xiang (2009) found a significant relationship between lower household income and problems with transportation among CSHCN. In Israel, Korn et al. (2007) found that low socioeconomic status has a direct correlation with high CRS non-compliance as 41% of non-CRS users CSHCN reported the high cost of a CRS or lack of funds as the leading challenge to CRS use. This is supported by local observational studies, not specifically on CSHCN, which found that affordability was the greatest challenge to CRS use followed by parents who felt that a seatbelt was a suitable alternative for their child's safety (Nagel, 2016; Puvanachandra et al., 2020). As many families are not able to purchase CRS, alternative solutions need to be pursued, such as the provision of low-cost/subsidised CRS or borrowing schemes to improve CRS accessibility (Puvanachandra et al., 2020).

Another factor influencing CRS usages in the current study was parental awareness of the South African legislation on vehicle safety restraints, with higher CRS usage found in parents knowing the legislation. The majority of parents showed good knowledge of seatbelt legislation, however, more than half did not know the current legislation on CRS use in South Africa. A recent literature review also found that parents who did not know the regulations are more likely to misuse CRS with their CSHCN (Downie et al., 2019). An observational study in Israel reported that 27% of parents did not know that a CRS is required for their CSHCN during transport, which could be due to the low parental education influencing safety knowledge as approximately 50% did not complete 12 years of education (Korn et al., 2007). Puvanachandra et al. (2020) supported this finding, in the South African context, as a higher level of education was significantly associated to CRS ownership.

The parent's perception of their child's sitting abilities impacted the use of CRS. Most parents who reported no sitting problems in their child did not transport their child in a CRS, even though the child is in the age and weight range to use a CRS as required by law. Further research could explore factors influencing parental decision making around CRS use.

3.6.5 Exploring the challenges faced by parents transporting children with special health care needs

The current study's results explored the challenges faced by parents when transporting their CSHCN in different modes of transport. Parents using public transport reported the CSHCN's poor sitting balance and lack of space within the vehicle to be their greatest challenge, whereas parents using private transport indicated that transporting the wheelchair and the unavailability of demarcated disability parking bays are the main transportation challenges. Some parents felt that transport services were unaffordable and could not accommodate their CSHCN's disability, resulting in them not travelling at all. A prospective study on CSHCN, used a series of standardised questionnaires to investigate the factors influencing their participation in daily activities and found similar accessibility challenges during transportation such as lack of adequate parking and poor accessibility to public transport (Forsyth et al., 2007). Accessible transportation services remain a global challenge for CSHCN, which can be addressed through legislation and improved infrastructure.

A third of the parents in the current study reported the need for their CSHCN to be supervised throughout the journey. This study did not focus on why there may be a need for supervision, however, during the focus group discussion participants mentioned that CSHCN at their schools require supervision because their behaviour may result in them unfastening their safety device or they may require immediate medical attention should they have a seizure. A recent systematic review describes the following parental concerns leading to the need for supervision; the CSHCN distracting the driver, moving the restraint into an unsafe position, freeing themselves from the CRS, negatively affecting others in the vehicle and the CSHCN opening the door of a moving vehicle (Downie et al., 2019). The need for supervision throughout the journey can influence the mode of transportation utilised additional passengers may be required to supervise the CSHCN while the driver remains focused on the road.

3.7   Study Limitations

While this study gained insight into how CHSCN are transported there were several limitations to using this study approach. This study population only included school-going CSHCN in an urban setting. It did not include parents and CSHCN who cannot access services such as education and healthcare, which might be due to transportation challenges. Another limitation is the potential sampling bias due to the size of the schools, resulting in one of the schools representing half of the sample population. In addition, there was possible response bias as data was not collected from families that did not respond to the questionnaire. These families may have had additional insights into transporting CSHCN.

It is widely accepted that questionnaires which are too long, can result in poorer response rates as respondents are less motivated to complete the form (Adams & Cox, 2008).

Questionnaires only allow for self-reported information to be gathered. Some of this information, such as CRS usage, household income or knowledge of legislation allows for response bias from the parents as they may have given a response that they feel the researcher wanted to hear (West, 2019). Due to possible overreporting, it would be beneficial to compare reported values to observed values in terms of seatbelt and CRS usage and misuse.

Despite thorough content validity through focus groups and suitable translation of the questionnaire, it cannot be excluded that parents may have misinterpreted complex questions (Adams & Cox, 2008). In addition, the translated questionnaire should have been translated back into English to prevent errors. Questionnaires do not promote an in-depth understanding of participant circumstance (West, 2019), suggesting that conducting interviews with open-ended questions may result in a richer understanding of the reason for participant behaviour. Other valuable information about transporting CSHCN that should be explored in future research includes whether mode of transport affect the type of CRS used, what design aspects of a CRS are important to parents, their perceptions about when to stop using a CRS and falls and accidents during transporting CSCHN.

Despite the questionnaire being contextually relevant in the South African setting, it was at times difficult to compare to other studies because the transport systems and the modes of transport are different. Private vehicle ownership is much higher in other countries (Falkmer & Gregersen, 2001; Wheeler, Yang & Xiang, 2009), whereas the current study's sample population made greater use of public transport due to the lack of private vehicles. As a result, this study focused on the limitations and challenges experienced during transportation, whereas other studies investigated in more detail the different travel destinations and their experiences (Falkmer & Gregersen, 2001; Wheeler, Yang & Xiang, 2009).

3.8 Conclusion

This study found that the modes of transport used by CSHCN vary depending on duration and distance travelled. Two thirds of the participants relied on school transport for their daily commute to school. When travelling outside of school hours for shorter distances, walking and using private transport had similar frequencies and private transport was used twice as often as public transport in longer distances. It also found that although parents report reasonable adherence to restraint and CRS use in vehicles, it is lower than surveys internationally, and observational studies suggest that these values may be overreported. Both historical and current use of CRS amongst CSHCN is lower than expected and it is important that the challenges such as knowledge of legislation, affordability and accessibility be addressed in ways that are appropriate to the South African context. In-depth interviews with parents would further explore the challenges to purchasing a CRS. Observational studies could be useful to gather more reliable information about CRS adherence and the implementation strategies to ensure more CSHCN are suitably restrained in the vehicle. As well as a naturalistic driving study will give more light on in-vehicle posture and behaviours of CSHCN.

Most parents did not know the current legislation on CRS which could impact CRS usage, this highlights the need for national campaigns to promote and educate citizens on road safety, including CRS legislation and the application in CSHCN. It is imperative to implement strategies to improve parental knowledge on seatbelt and CRS legislation via community education and caregiver training. As South Africa has low average monthly household income and poor private vehicle ownership, it may also be beneficial for government to regulate the availability of seatbelts and the procurement and use of CRS on public transport. Families who are unable to afford the high cost of specialised CRS for their CSHCN may need to seek funding from government or private charity organisations.

At the same time, it is also important to address the daily challenges and find feasible solutions to improve the safety, accessibility and utilisation of both public and private transport among CSHCN. Suggestions include universal access to all public transport systems, availability of specialised restraints, additional demarcated parking bays for persons with disabilities and construction of extra stops on public transport to reduce the walking distance to access the public transport.

Chapter 4: Effectiveness of currently available car restraint systems to maintain correct seating position during transportation for children with special health care needs

4.1 Introduction

A CRS only provides the desired protection when the child is correctly positioned, in particular their head should remain supported in the CRS throughout the journey (Andersson, Bohman & Osvalder, 2010). Whilst standard CRS used for transportation may provide adequate protection for many children, some CSHCN will need a CRS that provides additional postural support or is appropriate for children beyond the 36kg weight limit (O'Neil & Hoffman, 2019). Even CSHCN meeting the anthropometric requirements of standard CRS may require specialised CRS (McIntosh, Lindner & Suratno, 2013), to maximise postural support and accommodate deformities and medical equipment (Downie et al., 2019). Certain medical conditions may require flexibility of the seating posture or an adjustable harness to thread through medical devices (Brinkey, Manary & Santioni, 2011; Lindner, 2011). In particular, for CSHCN with orthopaedic conditions, breathing difficulties, reduced head and trunk control or changes in muscle tone, standard CRS may be inappropriate (Baker et al., 2012). For example, CSHCN using appliances such as orthoses or suffering from spasticity often struggle to use a standard CRS due to their limited joint range of movement (Lindner, 2011).

International studies have found that the use of appropriate CRS for CSHCN range widely between 30%-74.3% (Downie et al., 2019). The South African National Road Traffic Regulation 213 states that infants and young children, under the age of three years, must be seated in an appropriate CRS while being transported in a vehicle; and that a child, up to the age of fourteen years, must use an appropriate CRS where available (Department of Transport, 2014). An observational survey in the Western Cape, South Africa, showed that the CRS usage is poor amongst children entering at a hospital gate (Kling, 2011). However, the effectiveness of various CRS on postural control in this population has not been investigated. There are various CRS designs available in South Africa either locally or internationally produced which have different features and a wide variance in cost (Shonaquip, 2017; Sitwell Technologies, 2017; Wheelwell, 2017). Therefore, a comparison of the effectiveness of different CRS designs in providing postural support for CSHCN would be valuable.

The aim of the study is to I) determine the postural support needs of CSHCN in the Western Cape and II) evaluate the effectiveness of various CRS to maintain postural control of CSHCN during transportation. To achieve this aim, the study was divided into 3 different stages;

1. Inter- and intra-rater reliability testing of the self-designed tool used for assessing sitting balance.
2. A pilot study to determine face validity and feasibility of the recruitment, screening and assessment procedures prior to the main research study. This enabled the researcher to gain insight into the challenges that may be encountered during data collection on a larger scale and an opportunity to refine self-designed tools and procedures.
3. The main CRS study which determined the postural support needs of CSHCN and the suitability of different CRS designs.

The specific objectives of the main CRS study were to:
- Determine the postural support needs of CSHCN with difficulties with their sitting balance using the level of sitting scale to categorise their sitting ability.
- Determine if the postural support needs of CSHCN with postural control problems are met by locally available CRS, by taking photographs and categorising the posture, based on the degree of postural deviation.
- Describe characteristics of CSHCN who require specialised CRS

4.3   Methodology

4.3.1   Study design and research setting

A cross-sectional and pre-post study design was used to examine the relationship between the degree of postural control and the amount of postural support required from the CRS during transportation for CSHCN. Cross-sectional studies are relatively fast and inexpensive to conduct and can be used to determine prevalence among a population (Setia, 2016). The CRS study comprised of a self-designed screening tool, which had undergone reliability testing, a standardised sitting balance assessment for CSHCN and an investigation of the postural support provided by different CRS. A description of the research setting can be found in Chapters 1.4.

4.3.2 Participants

4.3.2.1 Main CRS study

Our study comprised of participants from the three institutions whose eligibility was based
on the following criteria:

<u>Inclusion Criteria</u>

- CSHCN aged between 4 -18 years-old who are enrolled at any of the participating
  institutions.
- All CSHCN whose transportation questionnaire was completed and returned by their
  parent/guardian, allowing for comparison between study findings.

<u>Exclusion Criteria</u>

- If no age- and cognitive-appropriate consent or assent could be obtained from CSHCN,
  as described in Chapter 4.3.4.5.3.
- CSHCN whose parents indicated that they underwent surgery in the past six weeks,
  such as spinal fusion, muscle tendon lengthening, which may have altered sitting
  balance or reported unstable health condition in the past six weeks, such as hip
  dislocation or seizures.
- Following the screening, those with no difficulties with sitting balance were also
  excluded from further testing.

4.3.2.2 Pilot CRS study

Of the total study participants, CSHCN from four classes in the Foundation Phase at the SID
school chosen by convenience, were included in the pilot study. These learners did not
undergo retesting as no change was expected and the data obtained in the pilot study was
included in the main CRS study.

4.3.3 Assessment tools

The outcome measures used in the CRS study were a self-designed 'traffic light' tool to
screen for difficulties with sitting balance, the LSS to formally assess the sitting ability of
CSHCN and various CRS related tests to analysis seated posture and determine the
suitability of the CRS during transportation.

4.3.3.1 Traffic light screening tool

Screening identifies individuals in a population who have an increased chance of meeting
the criteria of a health condition or disease. Persons who are identified as having a higher

chance, go on for further testing to confirm the diagnosis (The Society of Radiographers, 2018). The self-designed traffic light screening tool, to assess the ability of CSHCN to sit independently, was based on a three category approach (Basel Committee on Banking Supervision, 1996): green (no difficulties with sitting balance), orange (some difficulties with sitting balance) and red (difficulties with sitting balance) (Appendix 7.9). The screening tool, conducted by physiotherapists or occupational therapists at each school, identified CSHCN who may have difficulty with sitting balance during transportation and would require further postural control testing.

To ensure reliability of the tool, three therapists from the SID school conducted the inter- and intra-rater reliability testing. The screening was performed at the school and used the therapist's current knowledge of the CSHCN sitting ability, thus reducing the need for additional testing. This easy to complete, time efficient, screening tool reduced the burden on the investigator, therapists and CSHCN by avoiding re-assessment and preventing unnecessary time out of the classroom for the participants. A COSMIN checklist can also be found in Appendix 7.9 to describe the measurement properties that were assessed.

4.3.3.2   *Level of sitting scale*

CSHCN identified as having any difficulty with sitting balance, based on the traffic light screening tool (orange or red), were categorised by assessing their postural control during sitting, by the investigator, using the LSS (Fife et al., 1991). This ordinal scale has eight levels which are based on the amount of support required to maintain the sitting position or the stability of CSHCN if able to sit independently and has been validated for children with neuromotor disorders (Field, Debra A. & Roxborough, 2012) (Appendix 7.11). After completion of the LSS all participants went on to complete the CRS related tests.

4.3.3.3   *Testing suitability of car restraint systems*

Participant height and weight were considered according to WHO guidelines (World Health Organisation, 2009) to determine the most appropriate fit for a CRS. Weight categories were aligned with the different groups of CRS sizes, and the height categories were total height estimates based on the trunk height limits of the CRS. Participant weight was measured standing on a bathroom scale or seated on a chair scale, and trunk and/or total height was measured with a soft tape measure. CSHCN must be within both the weight and height category limits to be tested in the CRS.

The directional forces experienced during driving were simulated by pushing a CRS, secured in a wheelchair by a seatbelt, over a wooden ramp. The wheelchair was fitted with a three-point seatbelt by custom designed clamps to fix it to the frame in order to install and secure the different CRS to the wheelchair during testing (Figure 13). A pre-determined course was outlined to ensure that each participant experienced upward, downward, turning and tilting movements (Figure 14), this course was amended during the pilot study (Chapter 4.4.3.2).

The duration of each test of the CRS testing procedure was recorded in seconds on the data collection form to ensure consistency between tests (Appendix 7.12). In addition, the acceleration was measured to ensure acceleration did not exceed normal vehicle acceleration. Measuring acceleration with a mobile phone has been used to in transportation research because it can continuously obtain sensor data and does not depend on external equipment (Wang, Chen & Ma, 2010). A mobile phone with the accelerometer application to measure maximum acceleration experienced during the test was secured to the wheelchair by means of a universal mobile phone mount, ensuring that the accelerometer did not undergo additional movements or vibrations throughout the test and was easily visible to the investigator and/or research assistant (Figure 13).

During the pilot study, the acceleration was measured and tested for normality to compare to the acceleration experienced in a vehicle ($0.1 - 2.4m/s^2$) (Mehar, 2013). Acceleration testing was ceased after the pilot study as data showed that the acceleration experienced during testing was within the range of that experienced in a vehicle (Chapter 4.4.3.3), as well as to streamline the testing procedure and reduce the burden on participants by a longer test durations.

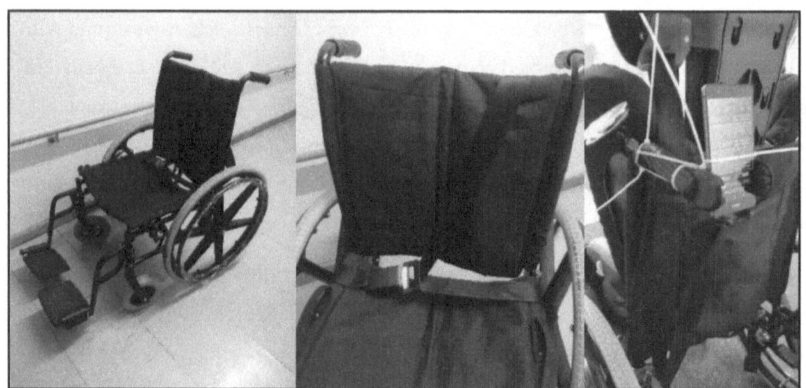

Figure 13: Mechanism of securing seatbelt and accelerometer to the wheelchair during testing

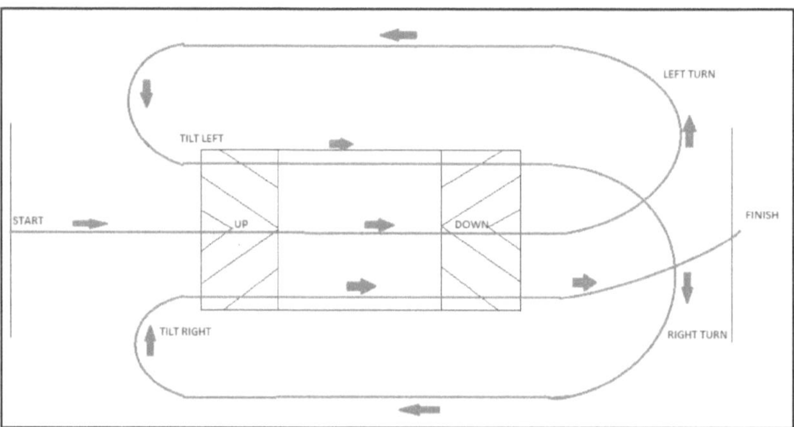

Figure 14: Preliminary course outline over ramp designed in the protocol

4.3.3.4   Seated posture analysis

Full-frame anterior and lateral photographs of the CSHCN's posture within the CRS were taken before and after the CRS test by the investigator at chest height to ensure standardisation. Each photograph of the CSHCN seating position was categorised based on the categorisation described by Andersson, Bohman and Osvalder (2010) for the degree of postural deviation of the head and torso from midline or the backrest of the CRS in the sagittal and frontal planes (Figure 15).

Head and torso positions were classified in both the sagittal and frontal planes and defined by the amount of deviation from neutral (Table 8 and Table 9). For example, in the AB position, in the sagittal plane, the child sits with the entire back against the backrest, while the head is upright. This position is comparable to the standard anthropomorphic test device (ATD) position, commonly known as the crash test dummy. In this position the torso remains against the backrest and within the side panels of the CRS, but there may be a slight deviation of the head in either plane.

A change in participant posture that resulted in a posture comparable to the ATD posture was acceptable whilst other postures were described as OOP. Acceptable postures were either 'AA' or 'AB' from the lateral view or 'aa' or 'ab' anteriorly. Acceptable postures of the head are either the 'A' or 'B' position and only 'A' for the torso. It was not possible to blind the data collector or analyser as the CSHCN were photographed in the CRS devices which are distinguishable through pictures.

Seated posture pre-test for each CSHCN was compared to their own post-test posture in each CRS. A comparison of the postural changes based on the categorised seating positions for each CSHCN, pre- and post-test, in different CRS was done during the analysis. The amount of change in posture between different seating categories is not equal, and as such only the presence of deterioration in posture was measured by the CRS performance. Performance of the different CRS was measured by the number of category changes in each of the four positions namely anterior and lateral head and torso positions. For example, if the pre-test postural analysis score for the anterior head position was (a) and changed to (c) post-test, then it would be noted as a deterioration. The sum of the number of deteriorations in all four positions was the CRS total performance score. If the CRS resulted in no change of posture for the head and torso in either the frontal or sagittal plane, then the total performance score would be 0. The worst performance for a CRS would be if there was a deterioration in all positions and a total performance score of 4. Total performance scores were compared between devices, and a correlated to CSHCN sitting ability.

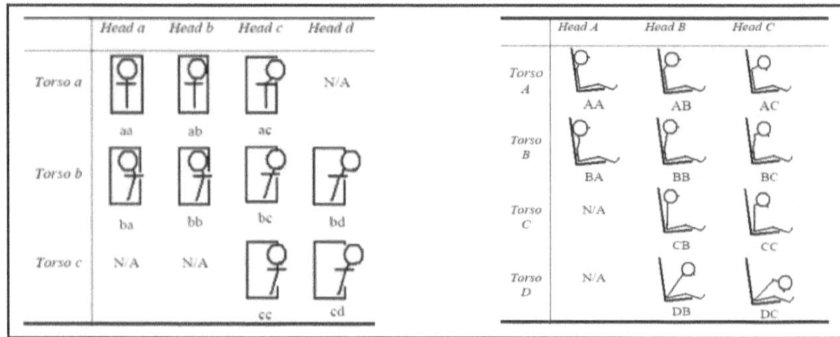

Figure 15: Categories in frontal plane (left) and sagittal plane (right) for analysis of the pre- and post-course lateral photographs Source: Andersson, Bohman and Osvalder (2010)

Table 8: Lateral torso and head deviations

Lateral torso positions	(A) the entire back including shoulders against the backrest
	(B) the entire back but not the shoulders against the backrest
	(C) child remains upright but no part of the back against the backrest
	(D) the torso is leaning forward without contact with the backrest
Lateral head positions	(A) head against the backrest
	(B) head upright relative to the torso
	(C) head leaning forward relative to the torso

Table 9: Anterior torso and head positions

Anterior torso positions	(a) the whole torso is within the backrest
	(b) one shoulder is outside the backrest, and
	(c) one shoulder and part of or the whole thorax is outside the backrest
Anterior head positions	(a) between the head side supports
	(b) resting against one of the head side supports
	(c) partly outside the head side supports
	(d) completely outside the head side supports

4.3.4   CRS study procedure

A diagrammatic reference to the study procedures can be found in Figure 16. The study procedure is described under the following phases: permissions and training, reliability testing of the screening tool, the pilot study and the main CRS study.

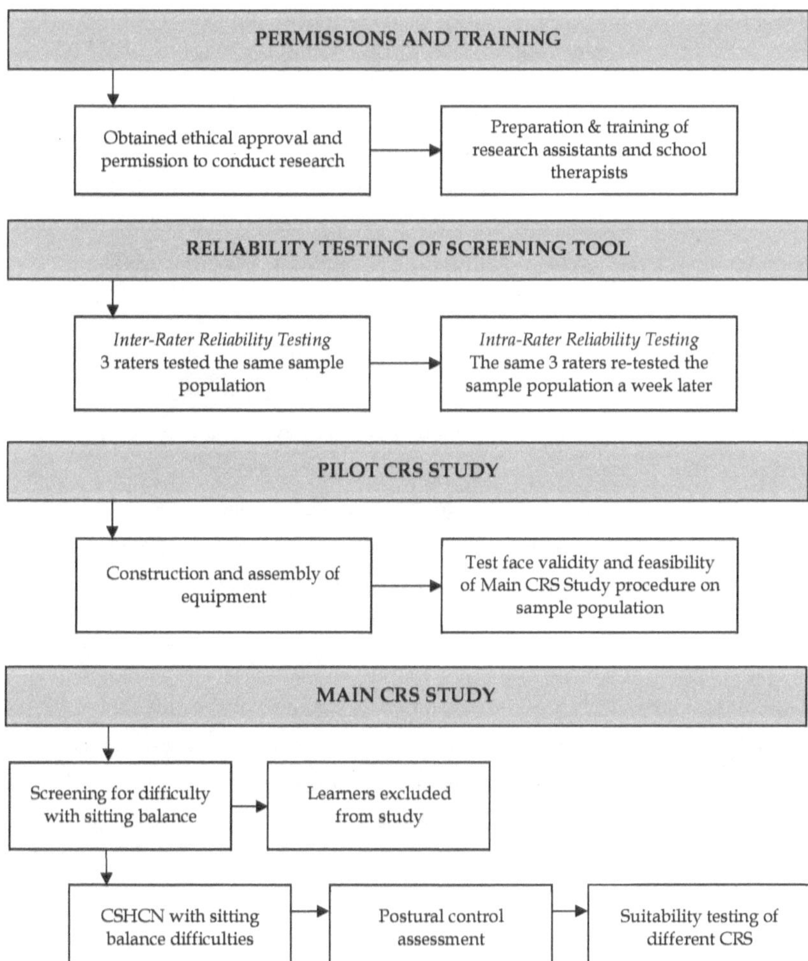

Figure 16: Flowchart of study procedures

4.3.4.1 *Permissions and training*

Ethical approval from HREC UCT (Appendix 7.3) and permission from the WCED (Appendix 7.5) to conduct research at the selected schools was obtained. Principals and/or governing bodies of participating schools were approached thereafter (Appendix 7.5), to 1) obtain permission to conduct research at their institution, 2) provide access to their learners, and 3) allow school therapists to assist in gathering information. Parental consent (Appendix 7.8) and CSHCN assent (Appendix 7.13), where possible, were obtained prior to enrolment.

4.3.4.2 *Preparation and training of research assistants and school therapists*

The school therapists, involved in screening the enrolled CSHCN suitable for further testing, were requested to consent to their participation in the study (Appendix 7.14). During an introductory meeting, the screening tool and a guide on how to complete it on the electronic class list was explained to the therapists (Appendix 7.9). Therapists discussed case examples until they were comfortable with the procedure and an opportunity to ask questions was given.

The research assistants were asked to sign a consent form with a confidentiality agreement to participate in the study (Appendix 7.15). Research assistants had experience in caring for and transferring CSHCN, however, the essentials of good ergonomics and handling were emphasised when showing them the correct ways to fasten the participants into each CRS. Research assistants aided during the CRS testing, to promote the safety of CSHCN by helping CSHCN who could not transfer independently, to transfer from their wheelchairs to the CRS secured in the testing wheelchair following which they were fastened securely in the CRS.

4.3.4.3 *Reliability testing of the screening tool*

The intra- and inter-rater reliability of the screening tool was evaluated by the three therapists at the SID school. Each therapist screened all the included learners from the Foundation Phase and results between therapists were compared to determine inter-rater reliability. Learners were retested one week later by the same therapists to determine intra-rater reliability for consistency of scores at the two time periods.

4.3.4.4 *Pilot CRS study*

The pilot CRS study commenced by constructing and assembling the equipment necessary for testing the suitability of the main CRS study. Thereafter it followed the main study procedure (Chapter 4.3.4.5) on the sample population to test face validity and feasibility of the procedure.

4.3.4.5 *Main CRS study*

CSHCN were screened for difficulties with sitting balance, provided assent for continued participation and then their sitting balance was assessed using the level of sitting scale. Height and weight were measured to assign the recommended CRS based on anthropometric fit, and observation of posture was done before and after completion of the test course to assess the suitability of different CRS.

4.3.4.5.1 *Screening for difficulties with sitting balance*
Occupational therapists or physiotherapists from each school used electronic alphabetical class lists and routinely collected data on the sitting ability of the CSHCN in their caseload to screen for postural control problems. These class lists did not include the names of CSHCN whose parents did not consent to participate.

4.3.4.5.2 *CSHCN excluded from the study*
Parents of CSHCN who were identified as having no problems with sitting balance were sent a letter informing them that their participation in the study was concluded along with an educational brochure on current best practice for CRS, seatbelt use for children and a list of local organisations supporting child road safety, validated by an expert in South African road safety (Appendix 7.16) (Mars, 2018).

4.3.4.5.3 *Assent by CSHCN for continued participation*
CSHCN identified as having difficulty with sitting balance provided assent to participate in the next phase of the study (Appendix 7.13). CSHCN assent was obtained at the school by the investigator prior to assessment and was adapted to the level of intellectual ability (either verbal or written). The procedure, risks and benefits were explained to the learners in an age- and cognitively appropriate manner.

Written assent was obtained from CSHCN at the school for specific learning disabilities. Verbal assent, where possible, was obtained from CSHCN at the school for severe intellectual disabilities. This was recorded by the person obtaining assent and signed by a witness acknowledging the CSHCN response. Witnesses for non-verbal communicating CSHCN were familiar communication partners.

Due to severe to profound intellectual disabilities at the CSPID school, it was not possible to obtain assent from these CSHCN, however, they received a verbal explanation of the procedure and were removed from the study if they showed signs of distress or discomfort such as moaning or facial grimace.

4.3.4.5.4 *Postural control assessment – Level of sitting scale*
To perform the LSS, the CSHCN was seated on a plinth without foot support and if necessary, supported by the investigator and/or the research assistant according to the categories of support described in the LSS (Appendix 7.11).

The height and weight limits of CRS testing groups were determined by recommendation set out by the WHO and CRS manufacturers (World Health Organisation, 2009). CSHCN were tested in a standard 3-point seatbelt only with no CRS and in up to three different CRS based on their anthropometric fit (Table 10). The standard CRS systems were either a Volvo forward-facing 5-point harness car seat or a booster seat secured with a 3-point belt. The locally produced specialised posture support seat, the IziPositioner from ShonaQuip came in three sizes and the internationally produced specialised CRS was the BeSafe IziUp Car Seat (Figure 17). Testing of each CRS took place on different days, to reduce potential postural fatigue.

Table 10: Categorisation of CRS devices according to WHO guidelines and anthropometric fit (World Health Organisation, 2009)

CRS Device		WHO Group	Weight Limit	Estimated Height Limit
(a) 3-point lap and diagonal seatbelt		Seatbelt	Unspecified	Unspecified
(b) 5-Point Harness Car Seat		1 & 2	25kg	95cm
(c) CRS Booster Seat		3	36kg	145cm
(d) Specialised CRS – Local	Small	N/A	Unspecified	Varied
	Medium	N/A	Unspecified	Varied
	Large	N/A	Unspecified	Varied
(e) Specialised CRS - International		2 & 3	36kg	145cm

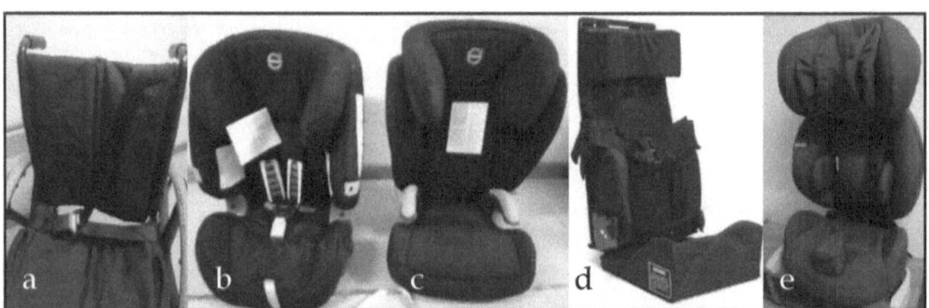

Figure 17: Devices tested for suitability (a) Seatbelt only, (b) 5-point harness CRS Car Seat, (c) CRS Booster Seat, (d) Specialised CRS - Local, (e) Specialised CRS - International

Each CRS was fitted on a wheelchair with a safety belt to allow the investigator to simulate vehicular movements by pushing the wheelchair over a self-designed course (Chapter 4.4.3.2). The course reproduced the directional forces of the vehicle and hence postural corrections of the CSHCN occurring during transport. The learner, therefore, experienced the different directional forces felt when travelling forward, uphill, downhill, tilting and turning to both sides as well as stopping. If the CSHCN could not be adequately secured in the CRS due to their severe physical limitations and it was therefore deemed unsafe to continue testing that CSHCN in that device.

There was a first aid trained personnel who ensured safety during the testing phase and a first aid kit was present in the eventuality that any injuries occurred during transfers between wheelchairs and CRS.

4.3.5 Data capturing and management

The data collection form (Appendix 7.12) collated data obtained during screening and the suitability testing before it was captured in a password protected Excel spreadsheet. Hard copies of the consent, assent and data forms were stored in a locked cupboard. All digital photographs were kept on a password protected hard drive. All information was coded, de-identified and imported to Statistica 13 for analysis (TIBCO Software Inc, 2018).

4.3.6 Data analysis

4.3.6.1 *Reliability testing of screening tool*

Screening tool results, from the pilot study sample, were collected as nominal data for the inter-rater and intra-rater reliability testing. Inter-rater reliability was calculated with Fleiss' Kappa and described by the level of agreement and a significant p-value of <0.05 (Table 11) (McHugh, 2012). Intra-rater reliability was calculated by the intraclass correlation coefficients with 95% confidence intervals (Table 12) (Koo & Li, 2016).

Table 11: Interpretation of Fleiss' Kappa results for inter-rater reliability as described by McHugh (2012)

Fleiss' Kappa (κ)	Level of agreement
≤ 0	No agreement
0.01–0.20	None to slight
0.21–0.40	Fair
0.41– 0.60	Moderate
0.61–0.80	Substantial
0.81–1.00	Almost perfect

Table 12: Interpretation of intraclass correlation coefficient for intra-rater reliability testing as described by Koo and Li (2016)

Intraclass correlation coefficient	Level of reliability
> 0.5	Poor
0.5 – 0.75	Moderate
0.75 – 0.9	Good
> 0.9	Excellent

Nominal and ordinal data such as age, weight, height and acceleration were tested for normality with the Shapiro-Wilk test (Ghasemi & Zahediasl, 2012), and analysed by calculating the mean and standard deviation (SD) or the median (IQR). The frequency distribution was examined for both nominal and ordinal data, such as diagnosis, weight and height categories, screening score, LSS and test duration.

Differences in non-parametric data, such as CRS types and test duration were tested with the Kruskal Wallis one-way ANOVA test. Postures and limb positions observed during testing, but not categorised by head and torso deviation from the midline are described alongside the photograph. All images are reproduced with consent and author permission is required for further reprinting.

Associations between categorical data, such as screening, LSS or performance score and posture deviation, were determined using Chi-square. A significant association was determined by a p-value of <0.05. Finally, the Spearman's Rho was used to determine the association between non-parametric results of LSS score and the CRS total performance score.

4.4   Results

The results from reliability testing of the screening tool, the pilot study and the CRS study are presented below according to research objectives.

4.4.1   Reliability testing of the screening tool

The procedure for using the screening tool was explained to the therapists and while they classified the sitting balance appropriately, there were troubleshooting errors when completing the form on the computer. The educational handout was improved to include screenshots of how to complete the class list and the electronic document was formatted so that any errors occurring from missing data was highlighted (Appendix 7.8). To determine the reliability of the screening tool 34 CSHCN, between 8 – 12 years old were screened a total of six times, twice by three different therapists.

4.4.1.1 *Inter-rater reliability*

Testing the self-designed screening tool, Table 13 indicates substantial agreement (κ=0.706) and an almost perfect agreement (κ=0.848 and κ=0.919) between raters with all of them showing statistical significance (p<.001), 95% CI (84.67, 99.93).

Table 13: Reliability between therapists scores – inter-rater reliability (n=34)

Therapist	Fleiss' Kappa (κ)	p
1	0.919	<.001
2	0.848	<.001
3	0.706	<.001

4.4.1.2 *Intra-rater reliability*

Interclass correlation coefficients and 95% confident intervals show good to excellent intra-rater reliability for all raters between their two scores (Table 14).

Table 14: Intra-rater reliability (n=33)

Therapist	ICC	95% Lower Confidence Interval	95% Upper Confidence Interval
1	0.903	0.816	0.950
2	0.890	0.792	0.944
3	0.913	0.834	0.956

4.4.2 Demographic characteristics of the pilot car restraint system study

Of the 34 CSHCN who were screened for difficulties with sitting balance, 22 did not present with any problems with their sitting balance and thus concluded their participation in the study. Of the remaining 12 CSHCN, only seven provided consent for further participation in the study.

The majority of the participants in the pilot study were males (57.4%; n=4) with a mean (SD) age of 10.43 (1.13) years (W=0.795; 0.046) (Figure 18). Height (W=0.884; p=0.272) had a mean (SD) of 137.71 (8.92) and weight (W=0.842; p=0.117) had a median (IQR) of 38.6kg (30.9 - 42.3). Five CSHCN were diagnosed with CP (71.43%). The rest of the data obtained for these CSHCN during the pilot study was combined with the main CRS study and will be presented in Chapter 4.4.4.

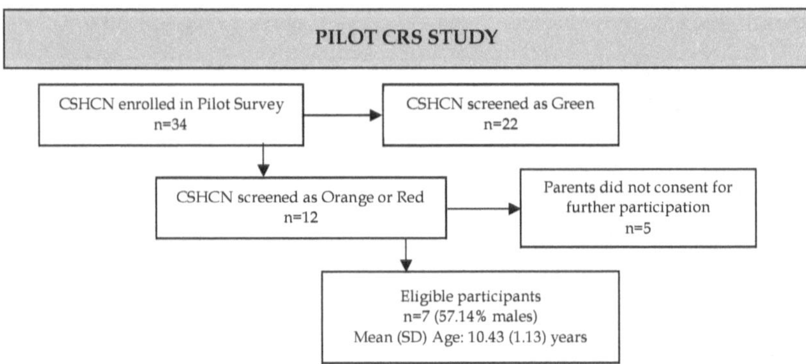

Figure 18: Flowchart of screening results and demographics of Pilot CRS study

4.4.3   Feasibility of the CRS study

The pilot study was a useful process to ensure that the procedures could be appropriately modified for effective and efficient data collection at the three institutions. This also confirmed that the study could be implemented in a practical manner to cause minimal disruption to the educational programs for the CSHCN. It was determined that the CRS study was feasible to conduct following minor amendments to the procedure.

Developing the screening tool, and manufacturing and assembling the equipment required for the CRS study, and implementing the procedure brought about various challenges and opportunities to streamline the protocol which will be discussed below.

4.4.3.1   *Ramp for CRS testing*

The three-piece wooden ramp, designed in the protocol, was developed to be compact and modular to fit in a vehicle for easy transportation between the different research sites. Prior to the construction, the ramp design was adjusted due to a concern of the incline being too steep. Instead of the initial 1:1 incline ratio intended, a slope with 1:2 was built. On the first test run, with the wheelchair only, it was evident that due to the length of the wheelchair wheelbase, the incline was still too steep as the front casters of the wheelchair hooked on the descend and was deemed unsafe.

Further research into ramps used for persons with disabilities was done to understand the recommendations for slope inclination. The National Building Regulations, considering the self-propelling wheelchair user, require the incline of ramps to be no steeper than 1:12

(South African National Standard, 2011). However, using this incline ratio would be impractical for transport purposes. The incline of three ramps at the SID school was measured to be 1:6, 1:9 and 1:10. These ramps are used by both self-propelling wheelchair users and wheelchair users who are propelled by others. The South African National Road Agency Road recommends that roads be constructed with an incline of between approximately 1:33 and 1:8, and a maximum gradient of 1:5 (The South African National Roads Agency, 2002).

Taking into consideration architectural and civil engineering recommendations, ramps in use with CSHCN at a special school, the need to mimic directional forces as well as needing to make the ramp as compact as possible for practical purposes, the new ramp was constructed with an incline of 1:5 (Figure 19). The height of the middle platform was also adjusted to ensure that an incline ratio of 1:5 remained during the lateral tilt when one wheel was on the ground and the other on the ramp platform.

Figure 19: Redesigned 3-piece modular ramp with an incline of 1:5

4.4.3.2   *General procedure*

To ensure therapists did not screen CSHCN of whom parents did not consent to participation in the study, the process of documenting the parents' consent to participate

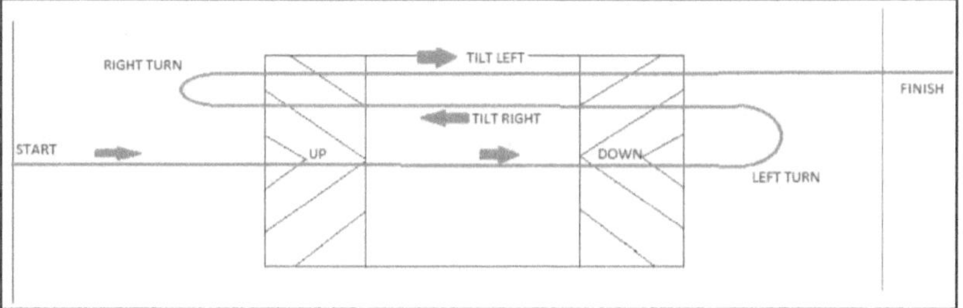

and the production of class lists for screening was improved via the use of spreadsheets. Instead of being taken to the data collection venue individually, participants were taken in small groups of up to five to streamline the data collection process and reduce the burden on the CSHCN by reducing class disruption as well as allowing them to familiarise themselves with what was expected during the test. Due to space restrictions in each venue the route over and around the ramp was adjusted based on the pilot study, ensuring feasibility of the protocol. Only one edge of the ramp and the number of turns was reduced. However, care was taken to ensure that all the elements that resulted in a directional force remained (Figure 20).

Figure 20: Amended course outline over ramp used during CRS testing

4.4.3.3   *Acceleration of CRS compared to on-road vehicles*
Data obtained during the pilot study confirmed that the acceleration experienced during the suitability testing is similar to that which is experienced in a car ($1.14 - 1.97 m/s^2$) (Table 15), as it did not exceed the maximum acceleration experienced in a standard car which ranges between $0.1 - 2.4 m/s^2$ (Mehar, 2013).

Table 15: Acceleration for Pilot CRS testing per device

Acceleration	n	Range m/s^2	Mean (SD) m/s^2
CRS - Booster seat	4	1,54 - 1,97	1,76 (0.18)
Specialised CRS - Local	4	1,54 - 1,81	1,65 (0.13)
Specialised CRS - International	3	1,22 - 1,85	1,60 (0.34)
Seatbelt only	6	1,14 - 1,97	1,67 (0.30)

4.4.4   Car restraint system study
All data obtained during the pilot and main CRS studies are presented in combination in the following sections: demographic characteristics, recommended CRS provision and postural support needs of CSHCN as well as the assessment of postural control to determine efficacy of different CRS.

4.4.4.1   *Demographic characteristics*
A total of 268 CSHCN who were screened for sitting balance difficulties by the therapists, of which 190 were found to have no problems hence not included in the study. Of the 78 found to have difficulties with sitting balance, three CSHCN left the research setting before data collection commenced, and two were excluded due to recent surgery. Of the resultant 73

participants, the majority were male (65.75%, n=48 (Figure 21). Their age was normally distributed (W= 0.961; p=0.022) with a mean (SD) of 11.56 (3.72) years (Figure 21 and Figure 22).

The majority of participants were diagnosed with CP (63.48% ; n=47) and between 95 - 145cm tall (67.12% ; n=49) (Table 16). Whilst the most common weight categories were between 18 - 36kg (43.84% ; n=32) and more than 36kg (39.73% ; n=29) (Table 16).

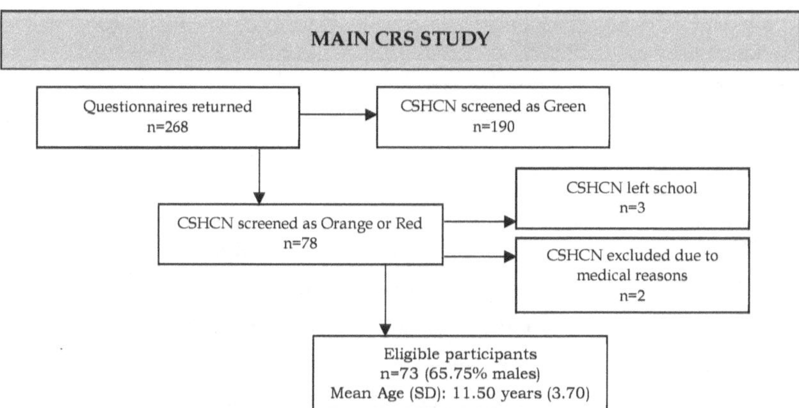

Figure 21: Flowchart of screening results and demographics of CRS study

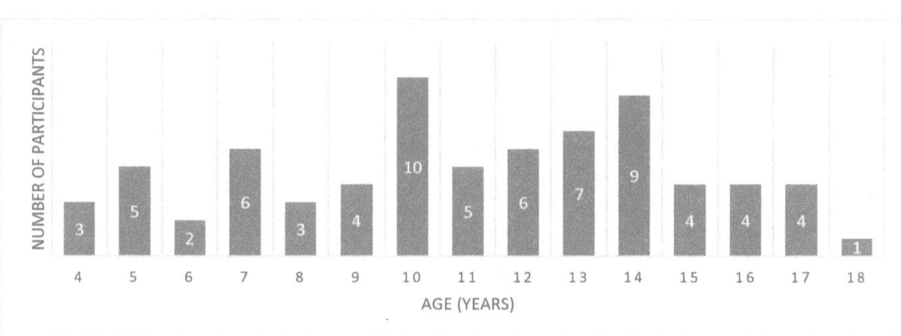

Figure 22: Age distribution in years (n=73)

Descriptive characteristic	Category	n (%)
Diagnosis	Cerebral Palsy	47 (64.38)
	Syndrome	5 (6.85)
	Spina Bifida	4 (5.48)
	Mild Motor Problem	4 (5.48)
	Muscular Dystrophy	3 (4.11)
	UMNL	1 (1.37)
	Unknown	12 (12.33)
Weight	<13kg	1 (1.37)
	13-18kg	11 (15.07)
	18-36kg	32 (43.84)
	>36kg	29 (39.73)
Height	<95cm	2 (2.74)
	95-145cm	49 (67.12)
	>145cm	22 (30.14)

4.4.4.2 Postural support needs of children with special health care needs

All the CSHCN participating in the study were screened to have difficulties in their sitting balance, 68.49% (n=50) of which some had difficulties with sitting balance (orange on screening) and 31.51% (n=23) had difficulties sitting in most circumstances (red on screening) (Figure 23). A significant association between the results of the LSS and the screening tool was found (X^2=53.75, p<0.001), with the majority of the participants not requiring support to maintain static sitting, being categorised as levels 5-8 of the LSS (78.08%, n=53) (Table 17 and Figure 24).

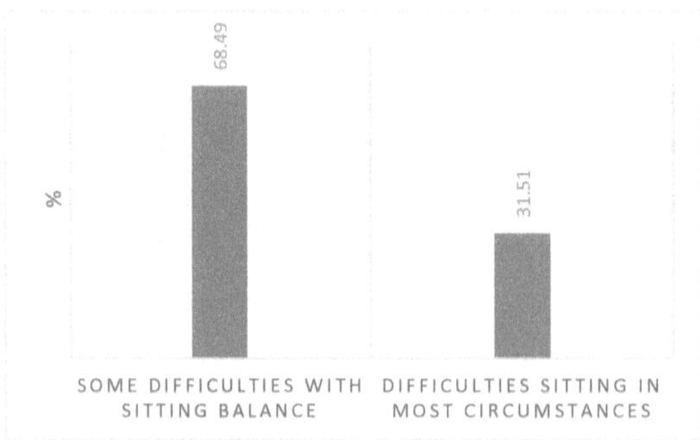

Figure 23: Screening of sitting balance by therapist using the self-designed screening tool in percent (n=73)

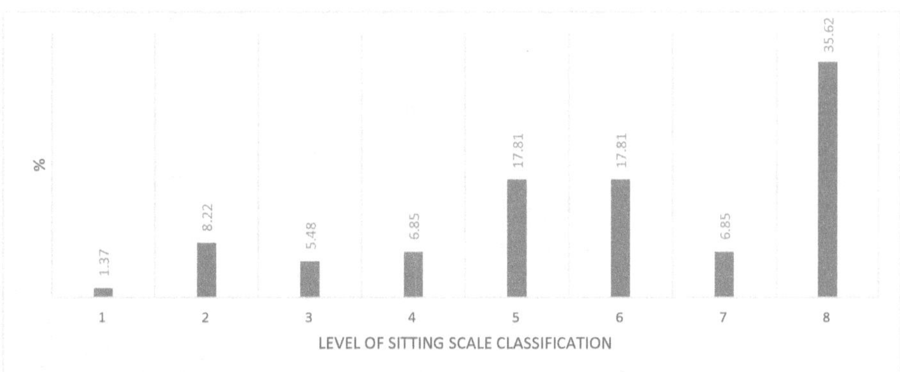

Figure 24: Distribution of Level of Sitting Scale in percent (n=73)

Table 17: Observed frequencies and two-way summary table for screening tool and LSS scores, based on n=73

Category	Subcategory	Some difficulties (Orange)	Difficulties most sitting (Red)	Row TOTAL	Statistical & p-value
LSS	1	0	1	1	X^2=53.75,
	2	0	6	6	p<0.001
	3	0	4	4	based on
	4	0	5	5	n=73
	5	7	6	13	(Yates
	6	12	1	13	correction)
	7	5	0	5	
	8	26	0	26	
	Totals	**50**	**23**	**73**	

4.4.4.3   Suitability testing of car restraint systems

The suitability and efficacy of the different CRS to provide postural control for CSHCN is described by categorising their head and torso posture pre- and post-test, analysing these postures per viewpoint and per CRS device as well as the overall performance score of each CRS device. Uncharacteristic postures that were not able to be quantified by the categorisation tool are also described in Chapter 4.4.4.3.2.

A total of 187 tests were conducted amongst the different CRS; CRS - Car seat (n=8), CRS – Booster seat (n=30), Specialised CRS – Local (n=49), Specialised CRS – International (n=33) and Seatbelt only (n=67). However, six tests could not be completed because the CSHCN could not be secured in the CRS due to their severe physical limitations, there were four participants in Seatbelt only, and one each in the CRS - Car seat and the Specialised CRS - Local.

Median (IQR) course duration for each CRS is tabulated in Table 18. A Kruskal Wallis one-way ANOVA test on the median test durations, between different CRS, was not significant ($H(4)$ = 4.45, p=0.349), showing that CRS types did not affect the test duration.

Table 18: Duration to complete the self-designed test course for each CRS, presented as median and IQR in seconds

	n	Shapiro-Wilk Test	Median (*in sec*)	IQR (*in sec*)
CRS - Car Seat	8	W=0.833; p=0.073	21.58	20.53-22.22
CRS - Booster Seat	30	W=0.660; p=4.307	22.59	20.78-23.96
Specialised CRS – Local	49	W=0.802; p<0.001	21.97	20.21-23.43
Specialised CRS – International	33	W=0.890; p=0.003	24.18	23.16-26.88
Seatbelt only	67	W=0.633; p=1.162	21.32	20.25-22.46

4.4.4.3.1 *Categorisation of pre- and post-test postures*

The OOP postures, a maximum number of four per test, one for each viewpoint, as explained in Chapter 4.3.3.4, are summarised below per CRS device. The seatbelt only demonstrated the largest proportion of OOP pre-test (13.43%, n=36), followed by the use of the Booster Seat (7.76%, n=9) (Table 19). These two devices remained the CRS with the largest proportion of OOP post-test and both had increased the number of OOP. The use of the Seatbelt only had 20.90% (n=56) of OOP and the Booster Seat had 18.33% (n=22) (Table 19). Examples of OOP postures in the anterior and lateral view for all devices can be seen

Figure 27 and Figure 28.

	Pre-Test n	Post-Test n	Pre-Test n (%)	Post-Test n (%)

CRS - Car Seat	9	8	2(5.56)	3(9.38)
CRS - Booster Seat	30	30	9(7.76)	22(18.33)
Specialised CRS – Local	50	49	10(5.10)	15(7.65)
Specialised CRS – International	33	33	8(6.25)	12(9.09)
Seatbelt only	71	67	36(13.43)	56(20.90)

position postures as a percentage of total tests per CRS device

Analysis of participants' posture pre-test shows that category A, with the head or torso fully supported and between the supports of the CRS (as described in Chapter 4.3.3.4), was the most common position for all CRS devices, from all viewpoints except for three: lateral head in the Car Seat; lateral head in the Booster Seat; and lateral head in the Specialised Seat - International. Post-test posture analysis shows that category A remains the most common position for all CRS devices, from all viewpoints except for the aforementioned views, as well as the anterior torso of the Seatbelt only. Results for each CRS, including associations between observed posture deviations pre- and post-test for each viewpoint of the different CRS designs, are presented in more detail in the below sections (

Figure 25 and Figure 26).

PRE-TEST POSTURE CATEGORSATION PER CRS TYPE

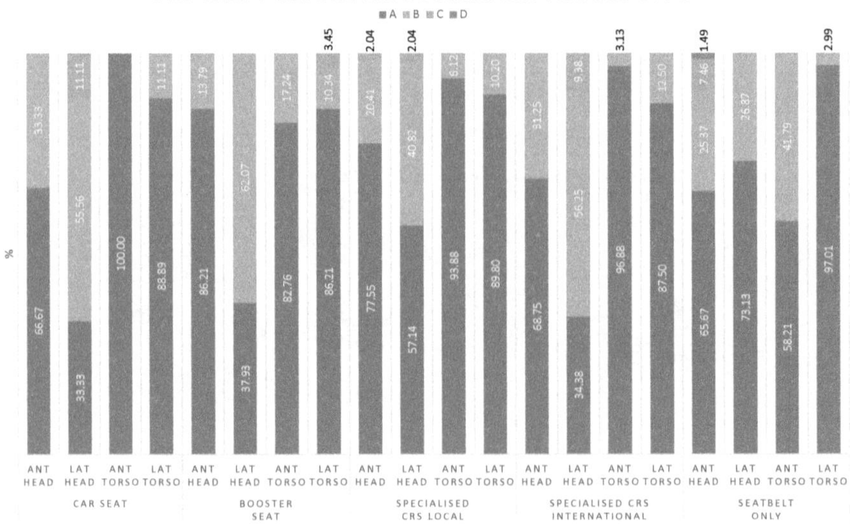

Figure 25: Pre-test categorisation of postures in all four viewpoints, in all CRS types in percent (CRS n=9, Booster seat n=30, Specialised CRS local n=50, Specialised CRS international n=33 and Seatbelt only n=68)

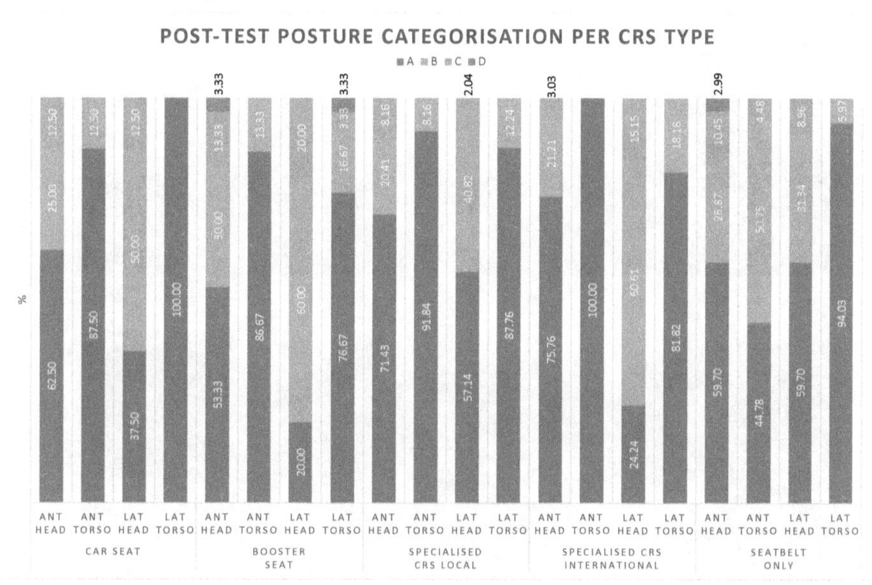

Figure 26: *Post-test categorisation of postures in all four viewpoints, in all CRS types in percent (CRS n=8, Booster seat n=30, Specialised CRS local n=49, Specialised CRS international n=33 and Seatbelt only n=67)*

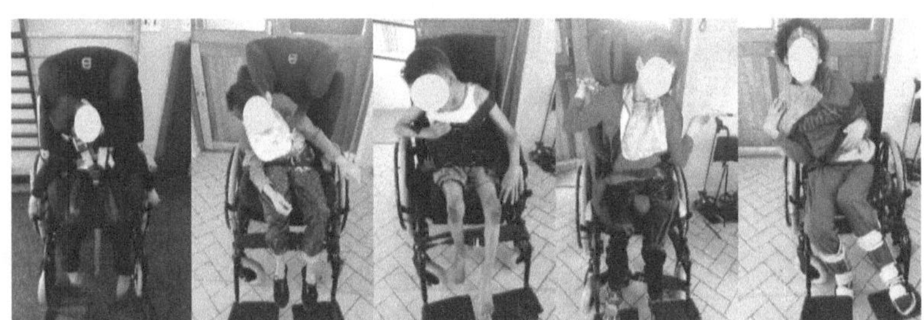

Figure 27: Examples of out-of-position postures in all CRS devices in the anterior view. (L-R) CRS Car Seat, CRS Booster Seat, Specialised CRS – Local, Specialised CRS – International & Seatbelt only

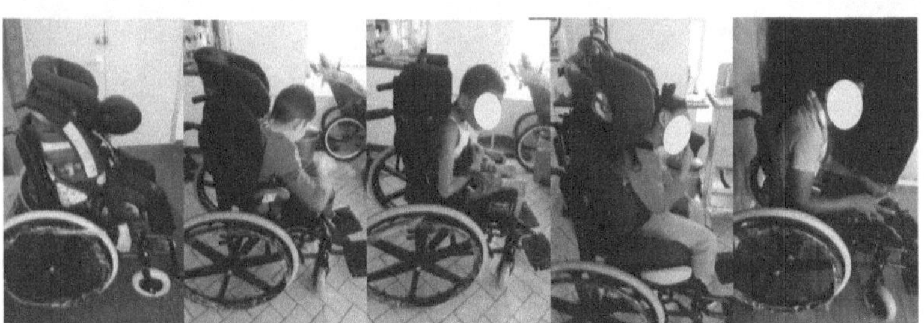

Figure 28: Examples of out-of-position postures in all CRS devices in the lateral view. (L-R) CRS Car Seat, CRS Booster Seat, Specialised CRS – Local, Specialised CRS – International & Seatbelt only

4.4.4.3.1.1 CRS car seat

Pre-test there was only 11.11% (n=1) OOP postures for both the lateral head and the lateral torso, however post-test the anterior head, lateral head and anterior torso all had 12.50% (n=1) OOP (Figure 29). No statistically significant associations were found between pre and post OOP postures.

4.4.4.3.1.2 CRS booster seat

Pre-test, OOP postures were only observed in the anterior and lateral torso (17.24%; n=5 and 13.79%; n=4 respectively). Post-test, OOP postures were seen in all four positions: lateral torso (23.33%; n=7), followed by lateral head (20.00%; n=6), anterior head (16.67%; n=5) and lastly anterior torso (13.33%; n= 4) (Figure 30). Post-test scores for the Booster Seat anterior head viewpoint were significantly associated with a deviated posture (X^2= 7.94, p=0.005) (Table 20). No other statistically significant associations were found between pre and post OOP postures.

Table 20: Observed frequencies and two-way summary table for posture deviations in the anterior viewpoint of the booster seat, based on n=60

Category	Subcategory	Pre-Test	Post-Test	Row TOTAL	Statistical & p-value
Booster seat Anterior Head	No Deviation	26	16	42	X^2=7.94,
	Deviation	4	14	18	p=0.005
	Totals	**30**	**30**	**60**	based on n=60

4.4.4.3.1.3 CRS local specialised seat

Only a few CSHCN were OOP in all postures pre-test: 10.20% (n=5) for lateral torso, 6.12% (n=3) for anterior torso and only 2.04% (n=1 for each) for both anterior and lateral head positions. There was only a slight increase in OOP post-test with lateral torso at 12.24% (n=6), followed by both anterior torso and anterior head at 8.16% (n=4), and finally the lateral head (2.04%; n=1) (Figure 31). No statistically significant associations were found between pre and post OOP postures.

4.4.4.3.1.4 CRS international specialised seat

The lateral torso (12.50%; n=4), lateral head (9.38%, n=3) and the anterior torso (3.13%; n=1) showed OOP postures pre-test. There were only slight changes post-test as OOP was the greatest for the lateral torso (18.18%; n=6) and the lateral head (15.15%; n=5), followed by the anterior head (3.03%; n=1) (Figure 32). Again, no statistically significant associations were found between pre and post OOP postures.

The Seatbelt only group showed the largest, single posture, OOP of 41.79% (n=28) for the anterior torso posture, followed by the anterior head (8.96%; n=6) and lateral torso (2.99%, n=2) positions. Post-test OOP postures included the anterior torso (55.22%; n=37), anterior head (13.34%; n=9), lateral head (8.96%; n=6) as well as the lateral torso (5.97%; n=4) (Figure 33). No statistically significant associations were found between pre and post OOP postures.

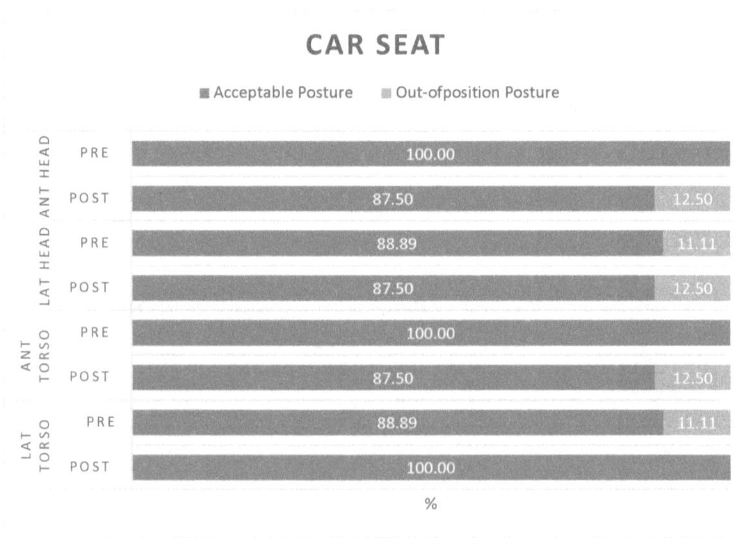

Figure 29: Acceptable and out-of-position postures pre- and post-test for CRS Car Seat (pre-test n=9; post-test n=8)

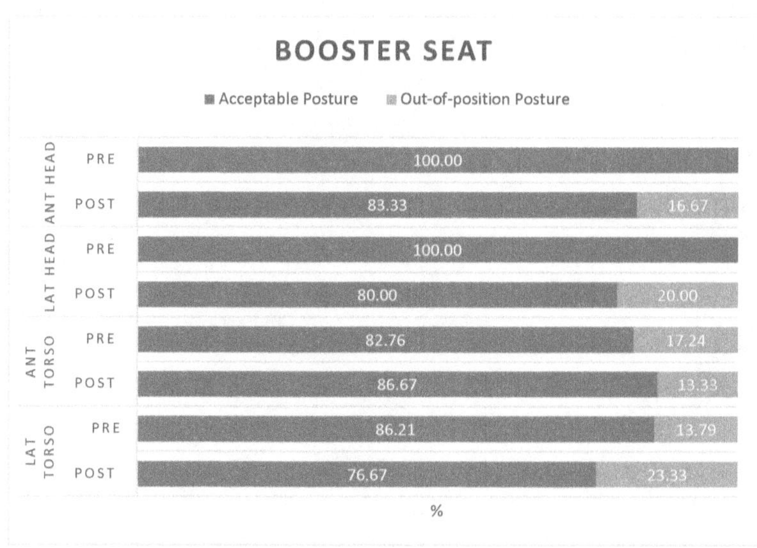

Figure 30: Acceptable and out-of-position postures pre- and post-test for CRS Booster seat (n=30)

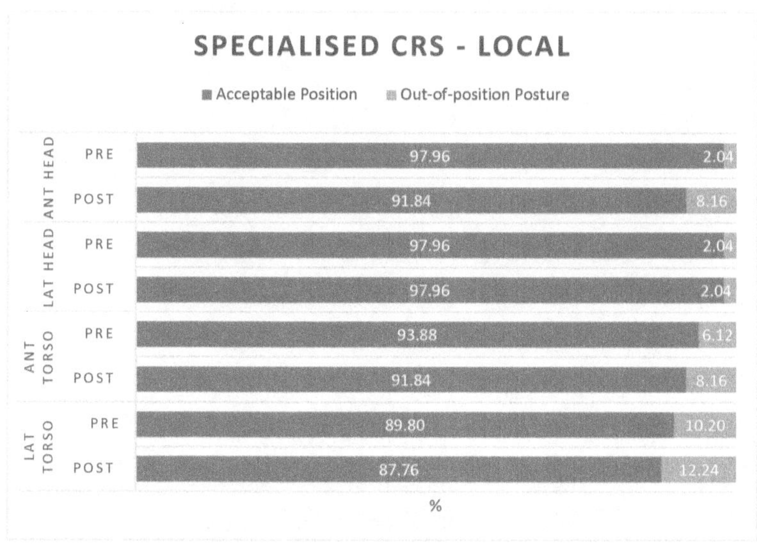

Figure 31: Acceptable and out-of-position postures pre- and post-test for CRS Local Specialised Seat postures (pre-test n=50; post-test n=49)

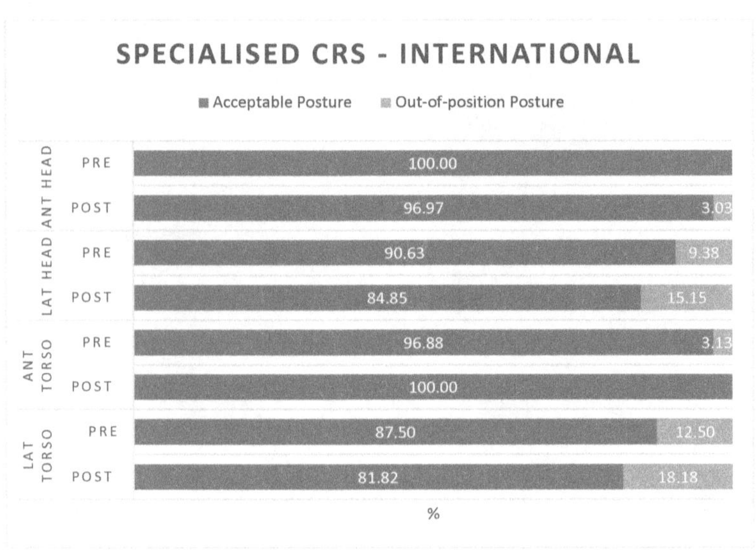

Figure 32: Acceptable and out-of-position postures pre- and post-test for CRS International Specialised Seat postures (n=33)

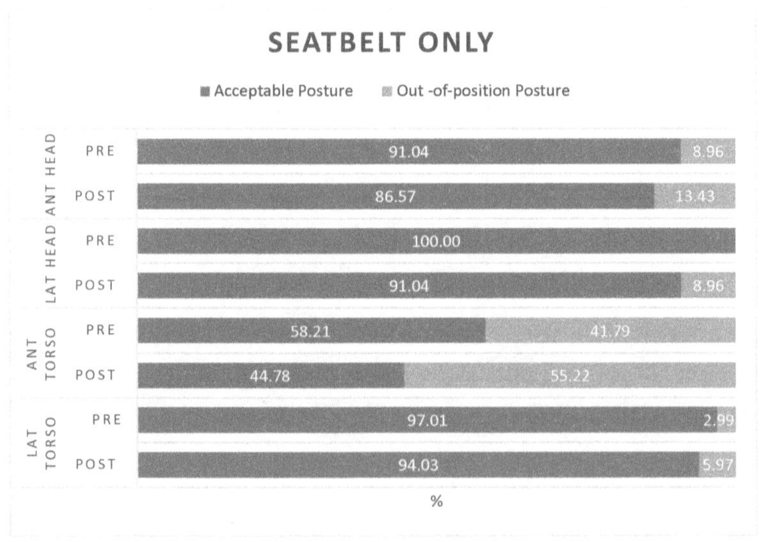

Figure 33: Acceptable and out-of-position postures pre- and post-test for Seatbelt only postures (pre-test n=68; post-test n=67)

The performance scores of each CRS device is first discussed as the performance score for each of the four viewpoints per device, then by the overall performance score for each device and lastly to associate the overall performance score with postural support assessment, the LSS.

The anterior head (46.67%; n=14) and the lateral head (43.33%, n=13) of the Booster Seat had the worst performance score and both viewpoints had majority of participants worsen their scores (Figure 34). The other CRS that lacked appropriate performance to maintain postural control were International Specialised CRS for the lateral head viewpoint (30.30%, n=10) and Seat Bely only for the anterior torso viewpoint (23.88%, n=16) (Figure 34). All devices performed well in the lateral torso viewpoint (<10.00%).

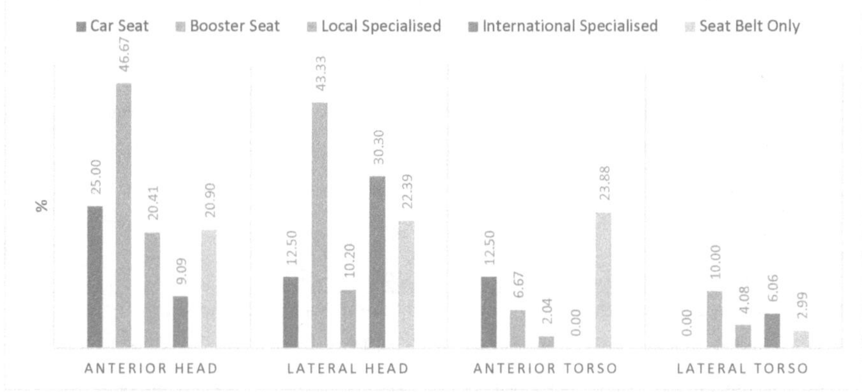

Figure 34: Percentage of performance scores indicating worsening of posture, per viewpoint, per CRS device (n=8, n=30, n=49, n=33, n=67)

The total number of worsening of postures were tallied and the CRS with the greatest number of participants with a poor performance were the Booster Seat (80.00%, n=24) and the Seatbelt only (55.23%, n=37) devices. The majority of participants tested with the Booster Seat had performance score of 1 (56.67%, n=17). The only CRS with performance scores of 3 were the Booster Seat (3.33%, n=1) and Local Specialised CRS (2.04%, n=1).

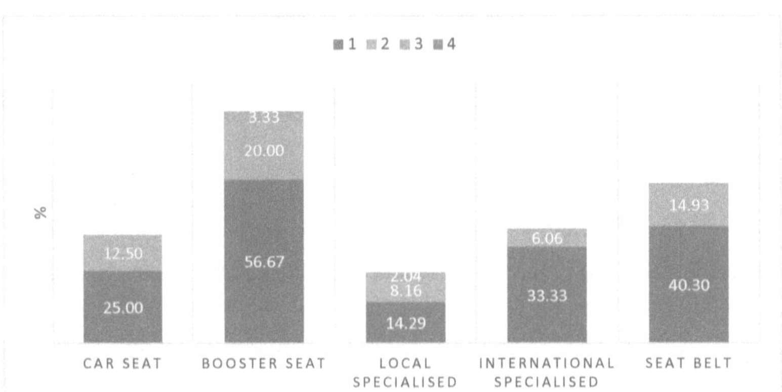

Figure 35: Total performance score (number of positions that each CRS had a worsening of postures) (n=8, n=30, n=49, n=33, n=67)

4.4.4.4   Characteristics of CHSCN who may require specialised CRS

4.4.4.4.1   Recommended car restraint system provision

Considering participant height and weight according to the WHO guidelines (World Health Organisation, 2009), 45.21% (n=33) of the participants should only need a seatbelt, the remaining participants should still use a CRS, either a booster seat (42.47%, n=31) or a car seat (12.33%, n=9).

4.4.4.4.2   Correlations between sitting balance and CRS performance score

There were no significant correlations between any of the total performance scores for each CRS device and the LSS score; Car Seat (r_s=-.50, p=0.173, n=9), Booster Seat (r_s=-.18, p=0.348, n=30), Local Specialised Seat (r_s=-.16, p=0.265, n=50), International Specialised Seat (r_s=.20, p=0.257, n=33) and Seatbelt only (r_s=-.12, p=0.341, n=68).

4.4.4.4.3   Noteworthy postures observed during testing

As the categorisation tool, described in Chapter 4.3.3.4, only considered deviations of the head and torso from the midline or the CRS backrest; some postures deviating from the standard ATD position went undocumented. For some participants their head and/or torso rotated in the transverse plane, and despite there being no resultant deviation of the head or torso from the midline, these abnormal postures might influence the risk of injury in an accident (Figure 36). A slouched position, with the upper back including the shoulders against the backrest, but the lower back and pelvis have no contact with the backrest was observed in all CRS. Some of these participants who presented with increased muscle tone or whose pelvis had moved forward so that their feet could be supported (Figure 37). Due to poor postural control of their head, and the height of backrest of the Seatbelt only CRS, some participants had no head support posteriorly and their heads extended backwards (Figure 38). Participants with increased tone, used their mass extension patterns to push themselves backward into the chair maintaining a good alignment, and possibly seeking foot support, however might not be sustainable during a full car journey (Figure 38). Limb position was not considered during the categorisation, however there were numerous participants who demonstrated inappropriate limb postures that would be an obstruction in-vehicle (Figure 39). One of the participants could not be adequately positioned in the CRS, hence unable to be tested, due to their limbs obstructing the wheels of the wheelchair (Figure 37). Finally, one participant felt it necessary to lift and their arm above his head and hold onto the headrest which he said was to "feel safer".

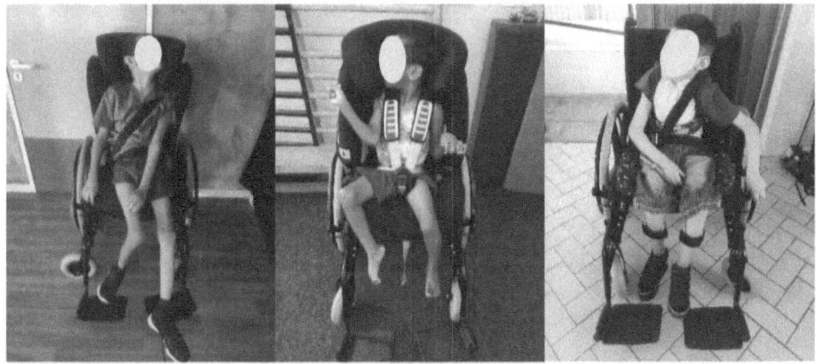

Figure 36: Examples of head and torso rotational deviations

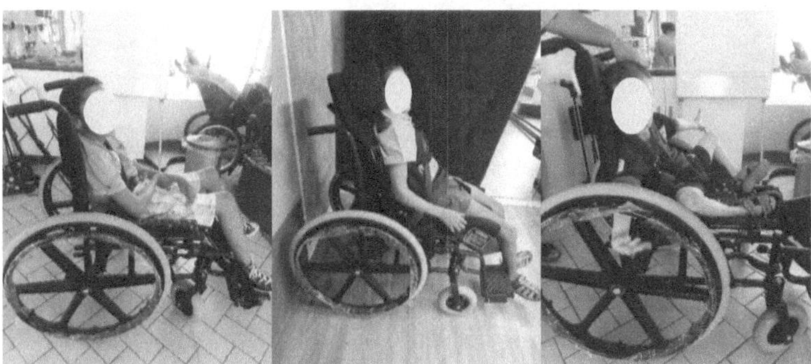

Figure 37: Examples of slouched posture

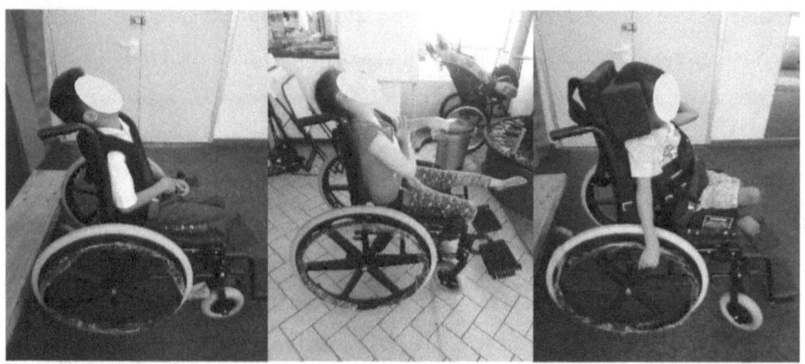

Figure 38: Examples of extension postures

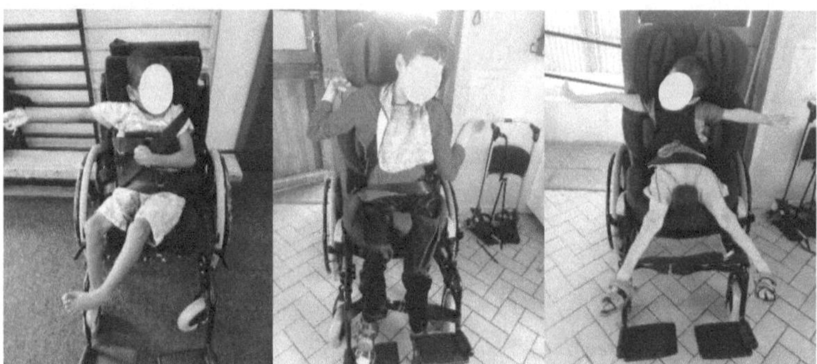

Figure 39: Examples of limb deviations

4.5 Discussion

To the best of the authors knowledge, this is the first study to determine the postural support needs of CSHCN in the Western Cape and whether these needs can be met by CRS during transportation. The results found that 21.92% (n=20) of CSHCN in this study required postural support to maintain sitting balance, determined by levels 1-4 on the LSS. The postural support needs of CSHCN are largely met by available CRS, however the worst performing devices were the Booster Seat and the Seatbelt only. These devices are intended for the older age groups of children and their design provides less postural support compared to the Car Seat or Specialised CRS. There was no significant association between the total performance scores or any of the CRS and the CHSCN LSS score.

Although the pilot study sample size was small (n=7), similar research on the posture of typically developing children during naturalistic driving studies have also had sample sizes under ten (Andersson, Bohman & Osvalder, 2010; Jakobsson et al., 2011; Osvalder et al., 2013). This study's pilot sample was therefore considered adequate for the feasibility testing, after which, pilot study data was merged with the main CRS study data.

Based on WHO recommendations the majority of this study's participants should be using either a Booster Seat or a Seatbelt only. According to their height (67.12% ; n=49 between 95 - 145cm tall), most study participants should still be making use of a Booster Seat, whereas according to weight the majority should either be using a Booster Seat (43.84% ; n=32 between 18 - 36kg) or a Seatbelt only (39.73% ; n=29 more than 36kg). Consequently, there were fewer investigations done with the CRS Car Seat designed for smaller and younger children. Further investigation may be needed for younger CSHCN requiring CRS Car Seats in particular. Naturalistic driving studies conducted with typically developing children investigated only one CRS group at a time and thus had smaller age range, 4 – 8 years old (Andersson, Bohman & Osvalder, 2010), 7 – 9 years old (Osvalder et al., 2013) and 8 - 13 years old (Jakobsson et al., 2011). This approach may be useful in future research if investigating one CRS group in greater detail.

4.5.1 Postural support needs of CSHCN with difficulties with their sitting balance

The screening tool was found to be reliable for inter- and intra-rater reliability as well the results were significantly associated to the LSS score. This easy to administer screening tool was effective in identifying CSHCN with difficulties with sitting balance from a larger population and requires only the therapist's knowledge of the CSHCN current sitting abilities without the need for re-evaluation. To be included in the study, all participants were screened to have at least some difficulties with sitting balance. There were 31.51% (n=23) of participants which were screened as having difficulties sitting in most circumstances, and 21.92% (n=20) required postural support to maintain sitting scoring between 1-4 on the LSS. There is a significant association between the results of the screening tool and the LSS (X^2=53.75, p<0.001), indicating that both of which may be useful in clinical practice to identity and classify CSHCN with difficulties with sitting balance. However, these tools evaluate different aspects of sitting balance as the screening tool considers sitting balance on surfaces of various stabilities, whereas the LSS tests sitting balance on a stable surface with no foot support. The LSS is useful for CSHCN with severe physical limitations as it describes the amount of support required if independent sitting is not possible. Further investigation on face and content validity of this screening tool may be useful before wider clinical implementation.

4.5.2   Are the postural support needs of CHSCN met by locally available CRS

Suitability testing of the CRS was performed on a simulated course, with a median duration of 21.32 to 24.18 seconds, rather than with naturalistic driving, which other studies have done with mean durations between 19 - 60 minutes (Andersson, Bohman & Osvalder, 2010; Charlton et al., 2010; Osvalder et al., 2013). The reason for this is that the simulated course procedure has a relatively low cost to implement, is safer and can be easily scaled for use with a larger population. Naturalistic driving studies are more expensive as specialised recording devices must be installed and fuel must be supplied for each trip. These studies provide a detailed analysis of posture throughout the test through video recording, furthermore, observing posture during naturalistic driving could enable the investigator to gain better insight in real life postural needs, see how behaviour guides posture, the presence of postural endurance and whether or not CSHCN can self-correct any deviated posture and return to the midline position. Despite the differences in testing procedure, using a simulated course is still a valuable comparison as a predictor for naturalistic driving studies.

There were CSHCN whose physical limitations were so severe that it was deemed unsafe to proceed with seven of the tests (Seatbelt only n=5; CRS Car Seat n=1 and Specialised CRS – Local n=1). For these participants, the CRS did not provide sufficient postural support to accommodate the severity of their physical disability by keeping their body and limbs in a neutral position safely contained within the CRS device to proceed with testing. Other studies investigating the posture associated to CRS use, did not include CSHCN in their sample population (Andersson, Bohman & Osvalder, 2010; Charlton et al., 2010; Jakobsson et al., 2011; Osvalder et al., 2013), so comparison to the current study is limited. An important area of investigation would be to examine in more detail how these CSHCN with severe physical limitations utilise transport and explore the design features of CRS that provide the postural support they require.

Virtual testing, is a useful method for extrapolating beyond currently existing test methods and crush test dummies, and has demonstrated that suboptimal belt positioning in a CRS due to postural changes is linked with OOP status and increases the risk of injury in an vehicle accident (van Rooij et al., 2005). OOP events, described as head, torso or limb deviations outside the protective margins of a CRS, result in a compromised position of the harness/seatbelt in most cases (Charlton et al., 2010). Inappropriate belt fit may also result in submarining, as the lap belt cuts into the abdomen while the pelvis slides underneath the belt (van Rooij et al., 2005). Furthermore, children which escaped the shoulder belt exceeded the head excursion limit by over 20cm in a simulation test (van Rooij et al., 2005), reaffirming the importance of maintaining proper belt/harness position. While belt position was not investigated in this study, it was observed that belt/harness position changed as postural changes occurred.

This study's findings on OOP postures in CRS demonstrated that the Seatbelt only (13.43%, n=36) and the Booster Seat (7.76%, n=9) had the largest proportion of OOP postures pre-test. Both devices remained the CRS with the largest proportion of OOP postures post-test (Seatbelt only had 20.90% ; n=56 the Booster Seat had 18.33% ; n=56). This may be because structurally these CRS offer the least torso support and provide no additional straps or harnesses. Observational studies, using similar posture categorisation as this study, investigating OOP postures amongst typically developing children during naturalistic driving in different CRS have found that there are substantial individual differences in children's' postures (Andersson, Bohman & Osvalder, 2010; Charlton et al., 2010; Jakobsson et al., 2011; Osvalder et al., 2013). Charlton et al. (2010) found that children ages 1 – 8 years were seated OOP for approximately 70% of the time, whereas the investigation by (Osvalder et al., 2013) found that children aged 7 – 9 years remained in the neutral position for 72 - 78% of the trip duration. In children between 3 – 6 years old, it was rare (4 – 11%) for the torso to lack contact with the seat's back (Andersson, Bohman & Osvalder, 2010). These extreme seating positions were initiated by specific activities like reaching for something or looking out the window and did not seem to be related to the seat design (Andersson, Bohman & Osvalder, 2010).

During naturalistic driving, children do not always sit as the ATD is positioned during crash tests (Brolin et al., 2015). Posture can be influenced by their state of alertness, awake and active children are more upright and forward leaning, whereas when children are resting and asleep the head leans all the way back or laterally on the headrest (Arbogast et al., 2016) Additionally, electronics can play a role in the shoulders moving forward off of the backrest and the head angled downward (Arbogast et al., 2016). As participating CSHCN in this study were not occupied with activities it is not likely that their posture deviations were a result of an intentional movement such as reaching or leaning. An aspect that was not explored in this study, and should be considered for future research, is the ability of the CSHCN to return to the upright position once postural deviation has taken place during naturalistic driving.

This study also investigated the prevalence of OOP postures post-test for each viewpoint to determine, more precisely, if CSHCN postural support needs are met by the CRS. No CRS prevented OOP posture in all CSHCN for the anterior and lateral head viewpoint. The Booster Seat, Specialised CRS - Local and the Seatbelt only had participants with OOP postures in all four viewpoints. A key observation in the current study is the lack of torso support for the majority of CSHCN in the anterior torso viewpoint of the Seatbelt Only CRS (55.22%, n=37), indicating that the use of a Seatbelt only does not provide adequate postural support for all CSHCN despite them meeting WHO anthropometric requirements. No significant association was found between the pre- and post-test scores of the Seatbelt only (X^2=2.14, p=0.144) which could be as a result of the large postural deviations pre-testing (41.79%, n=28) remained post-testing. However, there was a significant association between the pre- and post-test scores of the anterior head viewpoint of the Booster seat (X2= 7.94, p=0.005), indicating lateral head deviation.

In this study, the Booster seat had the most OOP postures in the lateral torso viewpoint (23.33% ; n=7). Another study, by Andersson, Bohman and Osvalder (2010), who compared two different Booster Seats, also found that the lateral head position was often deviated from normal. They speculated that this could be as a result of the temptation to look sideways over the head support. However, in the current study we did not find the CSHCN felt the urge to look around probably due to the lack of scenery and short course duration.

Overall performance scores, or the sum of the number of viewpoints where the CSHCN posture worsened post-test, attempt to indicate which CRS may provide better postural support considering all the viewpoints. When considering the performance score of the different viewpoints of each CRS device, the anterior head (46.67%; n=14) and the lateral head (43.33%, n=13) of the Booster Seat had the poorest performance and both viewpoints had majority of participants with a worsened overall performance score. All CRS devices performed well in the lateral torso viewpoint indicating that the torso was adequately restrained. Overall performance showed that the Booster Seat (80.00%, n=24) and the Seatbelt only (55.23%, n=37) had the majority of participants with a poor performance, thus providing the least amount of postural support to CSHCN. Poor performance of CRS to provide adequate postural support for CSHCN should encourage the design of new CRS to meet the unique seating needs of CSHCN.

Sitting postures obtained in the CRS are a result of the design specifications of the CRS (Jakobsson et al., 2011). A CRS with large side supports at the head and torso provides comfort and protection, but can only be achieved if the child's head is contained within the device (Andersson, Bohman & Osvalder, 2010). Using a Kinect sensor during naturalistic driving to observe head position of children between 1 - 8 years old, Arbogast et al. (2016) found that as the CRS type moved from more to less restraint, the range of head movements in both fore-aft and left-right movement increased. They also suggested that CRS position within the vehicle can influence the child's movement as children who were positioned in the centre seat showed the smallest range of head positions (Arbogast et al., 2016).

Uncategorised postures observed in the current study included head or torso rotation, abnormal limb placement, body extension and slouching. These postures were observed by the investigator amongst CSHCN with severe physical limitation, however, no association was found between LSS score and performance score for any of the CRS. Using photo observation to document OOP postures during long car drives, van Rooij, L et al., (2005) found that children between 1.5 - 3 year old tend to move around in their CRS, resulting in slanted and slouched postures which resulted in increased neck loads during virtual simulated load tests.

4.6   Study Limitations

Participants were selected from a convenience sample based on the survey study in Chapter 3: causing a possible sample bias as CSHCN not enrolled at one of the schools or who chose not to participate could provide further insight into postural support needs of CHSCN in CRS. Although the overall sample size was adequate data collected on the Car seat CRS was only for nine participants, as there were not many participants in the weight and height specifications of the device, which could skew the data. However, the Car seat CRS performed well overall as it is well designed with sufficient postural support.

It was not possible to blind the data collector and the analyser as participants would be photographed in the respective chairs which are identifiable through the pictures. There was a knowledge gap about the categorisation of postures at the time of protocol development, resulting in a lack of sensitivity for slouching. Future investigation on posture in CRS should include slouching postures as well as the relative belt/harness position if postural deviation occurs. Additionally, the speed of postural deviation from the midline, which was not investigated in this study, could be researched further as wider shoulders could result in the shoulders being positioned outside of the backrest quicker.

Due to the high variability in CHSCN size, diagnosis, severity of their disability as well as different CRS designs it was not possible to provide recommendations for which CRS would be most suitable for each CSHCN. Overall performance scores can give an indication of types of CRS designs that may provide the postural support required by CHSCN however, there is still the need for new CRS designs. The inclusion of a subjective performance analysis for each CRS device may have added value to the preferences of CSHCN to CRS design types.

This study only explored the postural support provided by two standard CRS, one locally produced and one internationally produced Specialised CRS, and Seatbelt only use. Although this study included the most affordable range of CRS design types, this selection bias may have prohibited other, more expensive, CRS designs for CSHCN from being included such as those imported by Sitwell Technologies (2017). These CRS may be less affordable, however their alternate design features such as rigid lateral torso supports, could result in improved postural support.

Other observational studies have observed postures in CRS during naturalistic driving by means of video analysis. These provide a clearer picture of the child's purposeful movement and quantified the postures for the duration of the journey. It was able to identify when children reached for objects and how they self-corrected back to the neutral position. Video analysis of CSHCN during naturalistic driving would provide insight into the daily behaviours of CSHCN in vehicles. Once it has been determined that a CSHCN can be safely and adequately supported in a CRS design type, future research should progress to continuous postural analysis during naturalistic driving for more in depth understanding and better comparison to research already conducted on typically developing children. Unfortunately, conducting research on CSHCN during naturalistic driving would require substantially more financial resources than what was needed for this study, and likely also resulting in small sample sizes. Future research could also examine if the postural support needs of CSHCN differ when using alternate modes of transport.

4.7   Conclusion

This study determined that the postural support needs of CHSCN are unique and depend on the child's anthropometry and the severity of their disability. More than half of participants should still be making use of a CRS according to their weight and height, but the currently available CRS designs may not provide the postural support needed for many CSHCN. This was observed by OOP postures or other postural deviations such as head and trunk rotation or limb position within the CRS which could be dangerous in the event of an accident.

The screening tool, found to be reliable for inter- and intra-rater reliability, may be a useful clinical tool for identifying CSHCN with difficulties with sitting balance for further investigations. There was a significant association between the screening tool and the LSS. The majority of participants were screened to have some difficulty with sitting balance and were also categorised as not requiring any support to maintain their static sitting balance on the LSS.

This study was not able to determine specific characteristics of CSHCN that require specialised CRS, as there was no association between the LSS and the overall performance score for any of the CRS devices. However, it was noted that the severity of their disability affected their ability to be tested.

The Seatbelt only and the Booster seat were the devices which showed the largest proportion of OOP postures. These devices provided the least amount of lateral support and did not offer additional harness support to secure the CSHCN other than the seatbelt. The viewpoints with the most deviations in posture were the lateral head of three CRS, resulting in the head moving anteriorly, and the anterior torso of the Seatbelt only, resulting in torso side flexion.

The Booster seat had the majority of participants worsen their posture overall and in both the anterior and lateral head viewpoints specifically. All devices performed well in the lateral torso viewpoint indicating that the torso remains against the backrest of the CRS. The Specialised CRS – Local had the best performance score overall, although this is only a positioning device not a safety seat. For effective implementation for CSHCN, CRS should be affordable, accessible, functional and accommodate growth and postural support needs. It would be valuable to observe CSHCN, of different ages and severity of disabilities, during naturalistic driving to gain a richer understanding of not only their postural control throughout the journey but also their behaviour.

Chapter 5: Book Conclusion

This book highlights the complex transportation needs of CSHCN in South Africa. The modes of transport used by CSHCN vary depending on duration and distance travelled and contextual challenges include knowledge of legislation, affordability and accessibility. Parents reported reasonable adherence to restraint and CRS use in vehicles, but previous studies suggest that these values may be overreported and further observational research is needed.

Postural support needs of CHSCN are unique and depend on the child's anthropometry and the severity of their disability. OOP postures include deviation of the head and torso from the midline or other deviations such as head and trunk rotation or limb position within the CRS. This study was not able to determine specific characteristics of CSHCN that require specialised CRS. It was however noted that the severity of their disability affected their ability to be tested.

The current CRS designs may be suitable for CSHCN who fit the manufacturers anthropometric requirements and whose physical limitations do not compromise the correct usage of the device. This study found that the head deviated more than the torso, but as this was a static assessment of posture we could not determine why the posture had changed.

5.1 Recommendations and Further Research

Based on the study conclusions, CSHCN transportation needs are individualised and unique, and vary depending on the family's access to different modes of transport. Practitioners prescribing and advising parents on CRS devices for the safe transportation of CSHCN should integrate thorough patient assessment and knowledge of CRS design specifications from the manufacturers. Policies should consider and accommodate for the challenges faced by CSHCN and their families in accessing, affording and utilising transport services. Advocacy and education programs should be combined with legislation enforcement to support improved implementation of CRS usage amongst all children, regardless of their disability status. For effective implementation for CSHCN, CRS should be affordable, accessible, functional and accommodate growth and postural support needs.

Further research could incorporate other specialised CRS designs, particularly ones that are suitable for CSHCN who have outgrown standard CRS sizes or those with severe physical limitations. Development of a categorisation tool that can quantify rotational postural deviations and a spectrum of limb positions would be beneficial in assessing the observed postures of CSHCN. It would also be valuable to observe CSHCN, of different ages and severity of disabilities, during naturalistic driving to gain a richer understanding of not only their postural control throughout the journey but also their behaviour.

Chapter 6: References

Adams, A. & Cox, A.L. 2008. Questionnaires, in-depth interviews and focus groups. *Research Methods for Human Computer Interaction*.17–34.

Andersson, M., Bohman, K. & Osvalder, A. 2010. Effect of booster seat design on children's choice of seating positions during naturalistic riding. *Annals of Advances in Automotive Medicine/Annual Scientific Conference*. Available: https://pubmed.ncbi.nlm.nih.gov/21050601/ [[19-03-2017]](54):171-180.

Angsupaisal, M., Maathuis, C.G.B. & Hadders-Algra, M. 2015. Adaptive seating systems in children with severe cerebral palsy across International Classification of Functioning, Disability and Health for Children and Youth version domains: a systematic review. *Developmental Medicine & Child Neurology*. 57(10):919-930. DOI:10.1111/dmcn.12762.

Arbogast, K.B., Kim, J., Loeb, H., Kuo, J., Koppel, S., Bohman, K. & Charlton, J.L. 2016. Naturalistic driving study of rear seat child occupants: Quantification of head position using a Kinect™ sensor. *Traffic injury prevention*. 17(sup1):168-174. DOI:https://doi.org/10.1080/15389588.2016.1194981

Arrive Alive. 2013. *Seatbelt wearing rates*. Arrive Alive National Fatal Accident Information Centre. Available: https://www.arrivealive.co.za/seatbelt-wearing-rates [Available: https://www.arrivealive.co.za/seatbelt-wearing-rates, 15-03-2019].

Arrive Alive. 2015. *Less than 7% of SA drivers put children in car seats*. Arrive Alive,. Available: https://www.aa.co.za/about/press-room/statements/less-than-7-of-sa-drivers-put-children-in-car-seats.html?Reference_ID=42537288 [Available: https://www.aa.co.za/about/press-room/statements/less-than-7-of-sa-drivers-put-children-in-car-seats.html?Reference_ID=42537288, 20-03-2017].

Arya, R., Antonisamy, B. & Garg, S.K. 2012. Sample Size Estimation in Prevalence Studies. *Indian Journal of Pediatrics*. 79. DOI:10.1007/s12098-012-0763-3.

Baker, A., Galvin, J., Vale, L. & Lindner, H. 2012. Restraint of children with additional needs in motor vehicles: Knowledge and challenges of paediatric occupational therapists in Victoria, Australia. *Australian Occupational Therapy Journal*. 59(1):17-22. DOI:10.1111/j.1440-1630.2011.00966.x.

Basel Committee on Banking Supervision. 1996. *Supervisory Framework for the Use of 'Backtesting' in Conjunction with the Internal Models Approach to Market Risk Capital Requirements*. Available: http://www.bis.org/publ/bcbs22.pdf [07-05-2017].

Brinkey, L., Manary, M. & Santioni, D. 2011. Development of an alternative five-point restraint harness to accommodate children with special healthcare needs in child safety seats. *Journal of pediatric rehabilitation medicine.* 4(4):289-300. DOI:10.3233/prm-2012-0192.

Brogren, E., Hadders-Algra, M. & Forssberg, H. 1998. Postural Control in Sitting Children with Cerebral Palsy. *Neuroscience & Biobehavioral Reviews.* 22(4):591-596. DOI:http://dx.doi.org/10.1016/S0149-7634(97)00049-3.

Brolin, K., Stockman, I., Andersson, M., Bohman, K., Gras, L.-L. & Jakobsson, L. 2015. Safety of children in cars: A review of biomechanical aspects and human body models. *IATSS Research.* 38(2):92-102. DOI:10.1016/j.iatssr.2014.09.001.

Brown, J., Wainohu, D., Aquilina, P., Suratno, B., Kelly, P. & Bilston, L.E. 2010. Accessory child safety harnesses: Do the risks outweigh the benefits? *Accident Analysis & Prevention.* 42(1):112-121. DOI:https://doi.org/10.1016/j.aap.2009.07.011.

Bull, M.J.M.D. 2008. Safe Transportation of Children with Special Healthcare Needs. *Pediatric Annals.* 37(9):624-631. Available: https://www-proquest-com.ezproxy.uct.ac.za/docview/217553654?accountid=14500.

Button, K.S., Ioannidis, J.P., Mokrysz, C., Nosek, B.A., Flint, J., Robinson, E.S. & Munafò, M.R. 2013. Power failure: why small sample size undermines the reliability of neuroscience. *Nature reviews neuroscience.* 14(5):365-376. DOI:10.1038/nrn3475.

Carlberg, E.B. & Hadders-Algra, M. 2005. Postural dysfunction in children with cerebral palsy: some implications for therapeutic guidance. *Neural Plasticity.* 12(2-3):221-228; discussion 263-272. DOI:10.1155/NP.2005.221.

Charlton, J., Koppel, S., Kopinathan, C. & Taranto, D. 2010. How Do Children Really Behave in Restraint Systems While Travelling in Cars? *Annals of Advances in Automotive Medicine.* 54:1-11. Available: https://www.researchgate.net/publication/289108515_How_do_children_really_behave_in_restraint_systems_while_travelling_in_cars [19-03-2017].

Child Accident Prevention Foundation of Southern Africa. 2013. *Buckle Up.*

Chung, J., Evans, J., Lee, C., Lee, J., Rabbani, Y., Roxborough, L. & Harris, S.R. 2008. Effectiveness of adaptive seating on sitting posture and postural control in children with Cerebral Palsy. *Pediatric Physical Therapy.* 20(4):303-317. DOI:10.1097/PEP.0b013e31818b7bdd.

Clay, C., van As, S.A.B., Hunter, K. & Peden, M. 2019. Latest results show urgent need to address child restraint use. *South African Medical Journal.* 109(2):66-66. DOI:10.7196/SAMJ.2019.v109i2.13831.

Connell, J., Carlton, J., Grundy, A., Taylor Buck, E., Keetharuth, A.D., Ricketts, T., Barkham, M., Robotham, D. et al. 2018. The importance of content and face validity in instrument development: lessons learnt from service users when developing the Recovering Quality of Life measure (ReQoL). *Quality of life research: an international journal of quality of life aspects of treatment, care and rehabilitation.* 27(7):1893-1902. DOI:10.1007/s11136-018-1847-y.

De Vries, E. & Reid, S. 2003. Do South African medical students of rural origin return to rural practice? *South African Medical Journal.* 93(10):789-793.

Department of Basic Education. 2015. *Report on the Implementation of Education White Paper 6 on Inclusive Education. An overview for the period: 2013-2015.*

Department of Basic Education. 2016. *Education Statistics in South Africa 2014.* Department of Basic Education.

Department of Transport. 2014. *Amendment of the National Road Traffic Regulations.*

Department of Transport. 2015. *National Learner Transport Policy.*

Department of Transport. 2020. *Public Transport Branch.* Available: https://www.transport.gov.za/web/department-of-transport/public-transport [2020, 04 May].

Downie, A., Chamberlain, A., Kuzminski, R., Vaz, S., Cuomo, B. & Falkmer, T. 2019. Road vehicle transportation of children with physical and behavioral disabilities: A literature review. *Scandinavian Journal of Occupational Therapy.* 27(5):309-322. DOI:10.1080/11038128.2019.1578408.

Easy Drive WC. 2016. Available: http://www.easydrivewc.co.za/index.html [Available: http://www.easydrivewc.co.za/index.html, 04-07-2017].

Economic Commission for Europe of the United Nations. 2011. *Regulation No 44 of the Economic Commission for Europe of the United Nations (UN/ECE) - Uniform provisions concerning the approval of restraining devices for child occupants of powerdriven vehicles ('Child Restraint Systems').* Available: http://eur-lex.europa.eu/LexUriServ/LexUriServ.do?uri=OJ:L:2011:233:0095:0210:en:PDF [03-07-2017]Official Journal of the European Union

Everly, J.S., Bull, M.J., Stroup, K.B., Goldsmith, J.J., Doll, J.P. & Russell, R. 1993. A Survey of Transportation Services for Children With Disabilities. *American Journal of Occupational Therapy.* 47(9):804-810. DOI:10.5014/ajot.47.9.804.

Falkmer, T. & Gregersen, N.P. 2001. A questionnaire-based survey on road vehicle travel habits of children with disabilities. *IATSS Research.* 25(1):32-41. DOI:https://doi.org/10.1016/S0386-1112(14)60004-2.

Falkmer, T. & Gregersen, N.P. 2002. Perceived risk among parents concerning the travel situation for children with disabilities. *Accident Analysis & Prevention.* 34(4):553-562. DOI:https://doi.org/10.1016/S0386-1112(14)60004-2.

Field, D. & Livingstone, R. 2013. Clinical tools that measure sitting posture, seated postural control or functional abilities in children with motor impairments: a systematic review. *Clinical rehabilitation.* 27(11):994-1004. DOI:10.1177/0269215513488122.

Field, D.A. & Roxborough, L.A. 2012. Validation of the relation between the type and amount of seating support provided and Level of Sitting Scale (LSS) scores for children with neuromotor disorders. *Developmental Neurorehabilitation.* 15(3):202-208. DOI:10.3109/17518423.2012.673177.

Fife, S.E., Roxborough, L.A., Armstrong, R.W., Harris, S.R., Gregson, J.L. & Field, D. 1991. Development of a clinical measure of postural control for assessment of adaptive seating in children with neuromotor disabilities. *Physical Therapy.* 71(12):981-993. DOI:10.1093/ptj/71.12.981

Forsyth, R., Colver, A., Alvanides, S., Woolley, M. & Lowe, M. 2007. Participation of young severely disabled children is influenced by their intrinsic impairments and environment. *Developmental Medicine & Child Neurology.* 49(5):345-349. DOI:https://doi.org/10.1111/j.1469-8749.2007.00345.x.

Foskett K. 2014. *Intellectual Disability in South Africa.* Available: http://www.includid.org.za/Downloads/South%20Africa%20and%20Intellectual%20Disability.pdf [03-07-2017].

Ghasemi, A. & Zahediasl, S. 2012. Normality tests for statistical analysis: a guide for non-statisticians. *International journal of endocrinology and metabolism.* 10(2):486-489. DOI:10.5812/ijem.3505.

Grant, R.W., Wexler, D.J., Watson, A.J., Lester, W.T., Cagliero, E., Campbell, E.G. & Nathan, D.M. 2007. How Doctors Choose Medications to Treat Type 2 Diabetes: A national survey of specialists and academic generalists. *Diabetes Care.* 30(6):1448-1453. DOI:10.2337/dc06-2499.

Green, E.M. & Nelham, R.L. 1991. Development of sitting ability, assessment of children with a motor handicap and prescription of appropriate seating systems. *Prosthet Orthot Int.* 15(3):203-216. DOI:10.3109/03093649109164290.

Harness, C. 2017. Available: http://www.crelling.com/ [Available: http://www.crelling.com/ [09-07-2017]].

Hayes, R.P., Fitzgerald, J.T. & Jacober, S.J. 2008. Primary care physician beliefs about insulin initiation in patients with type 2 diabetes. *International Journal of Clinical Practice.* 62(6):860-868. DOI:doi:10.1111/j.1742-1241.2008.01742.x.

Health Professions Council of South Africa. 2008. *HPCSA Guidelines for Good Practice in the Health Care Professions: Seeking Patients' Informed Consent: The Ethical Considerations. Booklet 13.* HPCSA Pretoria.

Healy, A., Ramsey, C. & Sexsmith, E. 1997. Postural support systems: their fabrication and functional use. *Developmental Medicine & Child Neurology.* 39(10):706-710. DOI:10.1111/j.1469-8749.1997.tb07368.x.

Hitge, G. & Vanderschuren, M. 2015. Comparison of travel time between private car and public transport in Cape Town. *Journal of the South African Institution of Civil Engineering.* 57(3):35-43.

Huang, P., Kallan, M.J., Neil, J., Bull, M.J., Blum, N.J. & Durbin, D.R. 2009. Children With Special Health Care Needs: Patterns of Safety Restraint Use, Seating Position, and Risk of Injury in Motor Vehicle Crashes. *Pediatrics.* 123(2):518. DOI:10.1542/peds.2008-0092.

Huang, P., Kallan, M.J., O'Neil, J., Bull, M.J., Blum, N.J. & Durbin, D.R. 2011. Children with Special Physical Health Care Needs: Restraint Use and Injury Risk in Motor Vehicle Crashes. *Maternal and Child Health Journal.* 15(7):949-954. DOI:10.1007/s10995-009-0539-1.

Jakobsson, L., Bohman, K., Stockman, I., Andersson, M. & Osvalder, A.-L. Eds. 2011. Older children's sitting postures when riding in the rear seat. 14-16.

Javouhey, E., Guérin, A.-C., Gadegbeku, B., Chiron, M. & Floret, D. 2006. Are restrained children under 15 years of age in cars as effectively protected as adults? *Archives of Disease in Childhood.* 91(4):304-308. DOI:10.1136/adc.2005.084756.

Jeske, C. 2016. Are Cars the New Cows? Changing Wealth Goods and Moral Economies in South Africa. *American Anthropologist.* 118(3):483-494.

Kling, J., Nicholls, T., Ntambeka, P. and Van As, A.B. 2011. Restraint use for child passengers in South Africa. *South African Medical Journal.* 101(9):146. DOI:10.7196/SAMJ.4737.

Knox, V. 2002. Evaluation of the Sitting Assessment Test for Children with Neuromotor Dysfunction as a Measurement Tool in Cerebral Palsy. *Physiotherapy.* 88(9):534-541. DOI:10.1016/S0031-9406(05)60136-8.

Koo, T.K. & Li, M.Y. 2016. A Guideline of Selecting and Reporting Intraclass Correlation Coefficients for Reliability Research. *Journal of Chiropractic Medicine.* 15(2):155-163. DOI:10.1016/j.jcm.2016.02.012.

Korn, T., Katz-Leurer, M., Meyer, S. & Gofin, R. 2007. How Children With Special Needs Travel With Their Parents: Observed Versus Reported Use of Vehicle Restraints. *Pediatrics.* 119(3):e637-e642. DOI:10.1542/peds.2006-1323.

Krosnick, J.A.V., D L 2018. Questionnaire design. In *The Palgrave handbook of survey research.* Springer. 439-455.

Leibbrandt, M., Finn, A. & Woolard, I. 2012. Describing and decomposing post-apartheid income inequality in South Africa. *Development Southern Africa.* 29(1):19-34. DOI:10.1080/0376835X.2012.645639.

Lindner, H. Ed. 2011. Child restraints for children with additional needs.

Locker, D.D. 2000. Telephone prompts improve questionnaire response rates. *Evidence-Based Dentistry.* 2(3):70-70. DOI:10.1038/sj.ebd.6400050.

Mars, P. 2018. *Personal Communication.* Personal Communication.

Mathews, C., Boon, H., Flisher, A. & Schaalma, H. 2006. Factors associated with teachers' implementation of HIV/AIDS education in secondary schools in Cape Town, South Africa. *AIDS care.* 18(4):388-397. DOI:10.1080/09540120500498203.

Maulik, P.K. & Darmstadt, G.L. 2007. Childhood disability in low- and middle-income countries: overview of screening, prevention, services, legislation, and epidemiology. *Pediatrics.* 120 Suppl 1:S1-55. DOI:10.1542/peds.2007-0043B.

McDonald, R., Surtees, R. & Wirz, S. 2004. The International Classification of Functioning, Disability and Health provides a model for adaptive seating interventions for children with cerebral palsy. *British Journal of Occupational Therapy.* 67(7):293-302. DOI:10.1177/030802260406700703.

McDougall, J. & Wright, V. 2009. The ICF-CY and Goal Attainment Scaling: Benefits of their combined use for pediatric practice. *Disability and Rehabilitation.* 31(16):1362-1372.

McHugh, M.L. 2012. Interrater reliability: the kappa statistic. *Biochemia medica: Biochemia medica.* 22(3):276-282. DOI:10.11613/BM.2012.031.

McIntosh, A., Lindner, H. & Suratno, B. Eds. 2013. Child Restraints for Children with Additional Needs.

McManus, V., Michelsen, S., Parkinson, K., Colver, A., Beckung, E., Pez, O. & Caravale, B. 2006. Discussion groups with parents of children with cerebral palsy in Europe designed to assist development of a relevant measure of environment. *Child: Care, Health and Development.* 32(2):185-192. DOI:10.1111/j.1365-2214.2006.00601.x

McPherson, M., Arango, P., Fox, H., Lauver, C., McManus, M., Newacheck, P.W., Perrin, J.M., Shonkoff, J.P. et al. 1998. A new definition of children with special health care needs. *Pediatrics.* 102(1):137-139. Available: https://portaldeboaspraticas.iff.fiocruz.br/biblioteca/a-new-definition-of-children-with-special-health-care-needs/.

Mehar, A., Chandra, S., Velmurugan, S. 2013. Speed and acceleration characteristics of different types of vehicles on multi-lane highways. *European Transport.* Available: https://www.semanticscholar.org/paper/Speed-and-acceleration-characteristics-of-different-Mehar-Chandra/7af13f02d6885ef5da90df151e813d2398359054 (55):1-12.

Morrongiello, B.A., Corbett, M. & Bellissimo, A. 2008. "Do as I say, not as I do": Family influences on children's safety and risk behaviors. *Health Psychology.* 27(4):498. DOI:10.1037/0278-6133.27.4.498.

Nagel, E. 2016. *Almost half of South African parents do not own car seats.* Available: https://blog.gumtree.co.za/almost-half-of-south-african-parents-do-not-own-car-seats/ [Available: https://blog.gumtree.co.za/almost-half-of-south-african-parents-do-not-own-car-seats/, 20-03-2017].

National Child Passenger Safety Board. 2012. *Federal Motor Vehicle Safety Standard No. 213.* Available: http://www.carseat.org/Legal/91_213sum.pdf [03-07-2017]

National Planning Commission. 2012. *National Development Plan 2030 Our Future-make it work.*

National Road Traffic Act, N.o. 1996. *National Road Traffic Act 93 of 1996.* South African Government.

Newby, A. 2012. *What is ISOFIX?* Available: https://www.whatcar.com/news/what-is-isofix/n11944 [Available: https://www.whatcar.com/news/what-is-isofix/n11944, 23-04-2019].

O'Neil, J., Yonkman, J., Talty, J. & Bull, M.J. 2009. Transporting children with special health care needs: comparing recommendations and practice. *Pediatrics*. 124(2):596-603. DOI:10.1542/peds.2008-1124.

O'Neil, J. & Hoffman, B.D. 2018. School Bus Transportation of Children With Special Health Care Needs. *Pediatrics*. 141(5):e20180513. DOI:10.1542/peds.2018-0513.

O'Neil, J. & Hoffman, B. 2019. Transporting Children With Special Health Care Needs. *Pediatrics*. 143(5):1-9. DOI:10.1542/peds.2019-0724.

Osvalder, A.-L., Hansson, I., Stockman, I., Carlsson, A., Bohman, K. & Jakobsson, L. Eds. 2013. Older children's sitting postures, behaviour and comfort experience during ride–A comparison between an Integrated Booster Cushion and a high-back booster.

Özkan, T., Puvanachandra, P., Lajunen, T., Hoe, C. & Hyder, A. 2012. The validity of self-reported seatbelt use in a country where levels of use are low. *Accident Analysis & Prevention*. 47:75-77. DOI:10.1016/j.aap.2012.01.015.

Parada, M.A., Cohn, L.D., Gonzalez, E., Byrd, T. & Cortes, M. 2001. The validity of self-reported seatbelt use: Hispanic and non-Hispanic drivers in El Paso. *Accident Analysis & Prevention*. 33(1):139-143. DOI:10.1016/S0001-4575(00)00012-9.

Parfitt, J. 2005. Questionnaire design and sampling. *Methods in human geography: A guide for students doing a research project*. 2. Available: https://eprints.soton.ac.uk/15698/.

Park, E.S., Park, C.I., Lee, H.J. & Cho, Y.S. 2001. The effect of electrical stimulation on the trunk control in young children with spastic diplegic cerebral palsy. *Journal of Korean medical science*. 16(3):347-350. DOI:10.3346/jkms.2001.16.3.347.

Petzäll, J. 1995. The design of entrances of taxis for elderly and disabled passengers. *Applied Ergonomics*. 26(5):343-352. DOI:http://dx.doi.org/10.1016/0003-6870(95)00051-8.

Pham, H.P., Eidem, A., Hansen, G., Nyquist, A., Vik, T. & Sæther, R. 2016. Validity and responsiveness of the trunk impairment scale and trunk control measurement scale in young individuals with cerebral palsy. *Physical & Occupational Therapy In Pediatrics*. 36(4):440-452. DOI:10.3109/01942638.2015.1127867

Puvanachandra, P., Janmohammed, A., Mtambeka, P., Prinsloo, M., As, S.V. & Peden, M.M. 2020. Affordability and availability of child restraints in an under-served population in South Africa. *International journal of environmental research and public health*. 17(6):1979. DOI:10.3390/ijerph17061979.

Reeve, K.N., Zurynski, Y.A., Elliott, E.J. & Bilston, L. 2007. Seatbelts and the law: how well do we protect Australian children? *Medical journal of Australia.* 186(12):635-638. DOI:10.5694/j.1326-5377.2007.tb01082.x.

Rosenbaum, P. 2003. Cerebral palsy: what parents and doctors want to know. *Bmj.* 326(7396):970-974. DOI:10.1136/bmj.326.7396.970.

Ryan, S. & Rigby, P. 2007. Toward development of effective custom child restraint systems in motor vehicles. *Assistive Technology.* 19(4):239-248. DOI:10.1080/10400435.2007.10131880.

Saether, R., Helbostad, J.L., Riphagen, II & Vik, T. 2013. Clinical tools to assess balance in children and adults with cerebral palsy: a systematic review. *Developmental Medicine & Child Neurology.* 55(11):988-999. DOI:10.1111/dmcn.12162.

Saloojee, G., Phohole, M., Saloojee, H. & C, I.J. 2007. Unmet health, welfare and educational needs of disabled children in an impoverished South African peri-urban township. *Child: Care, Health & Devlopment.* 33(3):230-235. DOI:10.1111/j.1365-2214.2006.00645.x.

Setia, M.S. 2016. Methodology series module 3: Cross-sectional studies. *Indian journal of dermatology.* 61(3):261. Available: https://www.e-ijd.org/text.asp?2016/61/3/261/182410.

Shonaquip. 2017. Available: http://shonaquip.co.za/ [Available: http://shonaquip.co.za/, 21-03-2017].

Simmons, A.E. 2018. *The Disadvantages of a Small Sample Size.* Available: https://sciencing.com/disadvantages-small-sample-size-8448532.html [Available: https://sciencing.com/disadvantages-small-sample-size-8448532.html, 15-10-2020].

Sitwell Technologies. 2017. Available: http://sitwell.co.za/ [Available: http://sitwell.co.za/, 03-07-2017].

South African National Standard. 2011. The application of the National Building Regulations Part S: Facilities for persons with disabilities. Available: https://www.google.com/url?sa=t&rct=j&q=&esrc=s&source=web&cd=&cad=rja&uact=8&ved=2ahUKEwiph9Xj-MjsAhUoneAKHTO9A5MQFjAAegQIBBAC&url=https%3A%2F%2Fwww.jica.go.jp%2Fsouthafrica%2Fenglish%2Factivities%2Fc8h0vm00005sup5w-att%2Factivities01_02.pdf&usg=AOvVaw0miZJUSdXwhC3-yFgSN5Dt.

South African Schools Act No. 84 of 1996

Standards New Zealand. 2013. *AS/NZS 4370:2013 Restraint of children with disabilities, or medical conditions in motor vehicles.* Available: https://shop.standards.govt.nz/catalog/4370:2013(AS%7CNZS)/scope [03-07-2017]

Statistics South Africa. 2014. *Census 2011: Profile of persons with disabilities in South Africa.* Available: http://www.statssa.gov.za/publications/Report-03-01-59/Report-03-01-592011.pdf [28-04-2017].

Statistics South Africa. 2015. *General Household Survey.* Available: https://www.statssa.gov.za/publications/P0318/P03182015.pdf [03-07-2017].

Statistics South Africa. 2017. *Labour market dynamics in South Africa,.* Available: http://www.statssa.gov.za/publications/Report-02-11-02/Report-02-11-022017.pdf.

Stout, J.D., Bull, M.J. & Stroup, K.B. 1989. Safe transportation for children with disabilities. *Am J Occup Ther.* 43(1):31-36. Available: https://www.ncbi.nlm.nih.gov/pubmed/2522277.

Tecklin, J.S. 2008. *Pediatric Physical Therapy.* Lippincott Williams & Wilkins. Available: https://books.google.co.za/books?id=YX8pztQHX0MC.

Teerds, S. & Cameron, T. Eds. 2015. Restraining children with disabilities or medical conditions safely.

The Society of Radiographers. 2018. *NHS Screening Programmes – Purpose of the programmes.*

The South African National Roads Agency. 2002. Geometric Design Guide Chapter 4 - Road Design Elements. Available: https://www.nra.co.za/content/GDGChapter4.pdf [08/02/2020].

TIBCO Software Inc. 2018. Statistica. version 13 ed.

van Rooij, L., Harkema, C., de Lange, R., de Jager, K., Bosch-Rekveldt, M. & Mooi, H. 2005. Child poses in child restraint systems related to injury potential: investigations by virtual testing. *TNO Science and Industry.* 05-0373. Available: https://www.researchgate.net/publication/238767337_CHILD_POSES_IN_CHILD_RESTRAINT_SYSTEMS_RELATED_TO_INJURY_POTENTIAL_INVESTIGATIONS_BY_VIRTUAL_TESTING [19-03-2017].

Walters, J. 2013. Overview of public transport policy developments in South Africa. *Research in Transportation Economics.* 39(1):34-45. DOI:https://doi.org/10.1016/j.retrec.2012.05.021.

Wang, S., Chen, C. & Ma, J. Eds. 2010. Accelerometer based transportation mode recognition on mobile phones. IEEE. 44-46.

West, J. 2019. *Data Collection.* Available: https://www.researchconnections.org/childcare/datamethods/survey.jsp [Available: https://www.researchconnections.org/childcare/datamethods/survey.jsp, 2020/10/07].

Wheeler, K., Yang, Y. & Xiang, H. 2009. Transportation use patterns of U.S. children and teenagers with disabilities. *Disability and Health Journal.* 2(3):158-164. DOI:https://doi.org/10.1016/j.dhjo.2009.03.003.

Wheelwell. 2017. *Services.* Available: http://www.wheelwell.co.za/services/ [Available: http://www.wheelwell.co.za/services/, 03-07-2017].

Whiteman, M.K., Langenberg, P., Kjerulff, K., McCarter, R. & Flaws, J.A. 2003. A randomized trial of incentives to improve response rates to a mailed women's health questionnaire. *Journal of Women's Health.* 12(8):821-828. DOI:10.1089/154099903322447783

World Bank Country and Lending Groups. Available: https://datahelpdesk.worldbank.org/knowledgebase/articles/906519-world-bank-country-and-lending-groups [Available: https://datahelpdesk.worldbank.org/knowledgebase/articles/906519-world-bank-country-and-lending-groups, 16-10-2020].

World Health Organisation. 2009. *Seat-belts and child restraint: a road safety manual for decision-makers and practitioners.* London.

World Health Organisation. 2015a. *Global Status Report on Road Safety.* Available: http://www.who.int/violence_injury_prevention/road_safety_status/2015/en/ [03-07-2017].

World Health Organisation. 2015b. *Global Health Observatory data repository: Seat-belt wearing rate Data by country.* Available: https://apps.who.int/gho/data/view.main.51416 [Available: https://apps.who.int/gho/data/view.main.51416 [05 May 2020]].

World Health Organisation. 2017. Draft Discussion Paper: Developing voluntary global performance targets for road safety risk factors and service delivery mechanisms. Available: https://www.who.int/violence_injury_prevention/road_traffic/WHODiscussionPaper-DevelopingVoluntaryGlobalPerformanceTargetsForRoadSafety_second_revision_22August2017.pdf?ua=1 [20-04-2020].

World Health Organisation. 2018a. *Global status report on road safety.* Available: https://www.who.int/violence_injury_prevention/road_safety_status/2018/en/ [12-04-2019].

World Health Organisation. 2018b. *Youth and road safety.* Available: https://www.who.int/violence_injury_prevention/publications/road_traffic/youth_roadsafety/en/ [13-04-2019].

World Health Organization. 2007. *International Classification of Functioning, Disability, and Health: Children & Youth Version: ICF-CY*. World Health Organization.

Zaza, S., Sleet, D.A., Thompson, R.S., Sosin, D.M., Bolen, J.C. & Services, T.F.o.C.P. 2001. Reviews of evidence regarding interventions to increase use of child safety seats. *American journal of preventive medicine*. 21(4):31-47. DOI:10.1016/S0749-3797(01)00377-4.

Chapter 7: Appendices

7.1 Amendments to self-designed transport questionnaire during focus group discussions

Table 21: Amendments to Self-Designed Transportation Questionnaire during Bel Porto School focus group discussion

Questionnaire Sub-section and concerns raised during focus group	Original questions	Adjustments made to questions Exclusions shown by strikethrough Inclusions indicated by […]
Understanding your child's disability Question 5 - Concerned that parents will lose focus and misinterpret the question if explanations are lengthy - Condense the answers by adding "My child" to the question instead of the answer - Clarify the intended surface when describing the "child's ability to sit independently"	Please describe your child's ability to sit independently a) My child can sit independently on any type of chair without using their hands to support them. They can also reach across their body to pick up something without falling over. b) My child can sit independently on most types of chairs but falls if they reach across the body. c) My child can sit independently on a chair with a backrest or must use their hands for support. d) My child can sit with some assistance in a well-supported chair with a backrest and armrests. e) My child needs a lot of support from the wheelchair or someone holding them in order to sit.	Please ~~describe~~ [choose which option best describes] your child's ability to sit ~~independently~~ [on a chair. My child…] a) ~~My child can~~ [Can] sit ~~independently~~ on any type of chair without using their hands to support them. They can also reach across their body to pick up something without falling over. b) ~~My child can~~ [Can] sit independently on most types of chairs but falls if they reach across the body. c) ~~My child can sit independently on a chair with~~ [Requires] a backrest or must use their hands for support. d) ~~My child can s~~ [S]its with ~~some~~ [the] assistance ~~in~~ [of] a well-supported chair with a backrest and armrests.

		e) ~~My child n~~ [N]eeds a lot of support from the wheelchair or someone holding them in order to sit.
Question 6 - Allow the option to tick more than one box if appropriate - Explain each medical term in layman's terms to ensure better understanding	Please tick the box if you feel your child has any of the following secondary complications: ☐ Spasticity in the muscles (mobile joint) ☐ Joint contractures (immobile joint) ☐ Scoliosis ☐ Hip dysplasia ☐ Loss of sensation	Please tick the box[(es)] if you feel your child has any of the following secondary complications: ☐ Spasticity in the muscles (~~mobile joint~~ [stiffness in the muscles of the arms and/or legs, but they can still be straightened]) ☐ Joint contractures (~~immobile joint~~ [stiff muscles stop joints from being straightened]) ☐ Scoliosis [(curved spine/back)] ☐ Hip dysplasia [(full or partial dislocation of the hip)] ☐ Loss of sensation [(unable to feel part of the body)]
Question 7 - Include an option to indicate that the child did not undergo surgery	In the last six weeks, has your child undergone any surgery? Please tick box if relevant or list surgery: ☐ Spinal fusion ☐ Muscle tendon lengthening ☐ Hip surgery ☐ Other: (Please describe:)________________	In the last six weeks, has your child undergone any surgery? Please tick box if relevant or list surgery: ☐ [None] ☐ Spinal fusion ☐ Muscle tendon lengthening ☐ Hip surgery Other: (Please describe:)________________
Transportation		

Question 8	How does your child travel to school?	How does your child travel to school?
- Further describe which type of public or private transport is used	a) Walk or pushed in wheelchair b) School combi (small capacity) c) School bus (large capacity – different style seats) d) Public transport (Train, Taxi, Travel Club, MyCiti/Golden Arrow bus) e) Private transport (Car, Combi, Bakkie)	a) Walk or pushed in wheelchair b) School combi (small capacity) c) School bus (large capacity – different style seats) d) Public transport (Train, Taxi, Travel Club, MyCiti/Golden Arrow bus) e) Private transport (Car, Combi, Bakkie) [If you answered d) or e) in Question 8. Please indicate what type of public transport or vehicle is used:________________________]
Question 9 - Split this question so that transport to near or far activities can be described separately if necessary. - Include a question allowing for learners who do not travel unless it is for school. - Allow a description for the type of public or private transport used - Include an option for the child not going out and an opportunity to describe these challenges	How does your child commute, most often, for other activities? (Church, shops, friends) a) Walk or pushed in wheelchair b) Public transport (Train, Taxi, MyCiti/Golden Arrow bus) c) Private transport (Car, Combi, Bakkie)	How does your child commute ~~most often, for other activities? (Church, shops, friends)~~ [to activities/places near to your home? (less than 5km or 15min drive) a) Walk or pushed in wheelchair b) Public transport (Train, Taxi, MyCiti/Golden Arrow bus) c) Private transport (Car, Combi, Bakkie) d) [My child does not go out because of the challenges faced] How does your child commute ~~most often, for other activities? (Church, shops, friends)~~ [to activities/places far from your home (more than 5km or 15min drive)]

		[If you answered b) or c) in Question 9 or 10. Please indicate what type of public transport or vehicle is used:________________________]
		[Only answer this question if you indicated "My child does not go out because of the challenges faced" in either question 9 or 10. Please tick which box(es) describe the challenges faced: a) It takes too long to get my child ready for a journey b) I cannot take my child's wheelchair with us in the vehicle c) I feel that it is unsafe to travel with my child d) My child and their wheelchair cannot be accommodated n the public transport e) I cannot afford the costs of a private taxi f) It is too difficult to get my child into the vehicle g) The public transport is too far from my house/destination and I cannot push the wheelchair that far]
Question 10 - Remove negativity of the phrase "not applicable" - Allow an explanation for why the wheelchair/pram is not transported with the child	How do you transport your child's wheelchair when commuting? a) Not applicable - my child does not have a wheelchair	How do you transport your child's wheelchair when commuting? a) ~~Not applicable - m~~[My] child does not have a wheelchair

- Wheelchair may be placed in other part of vehicle than just in the boot. Question 12 - Condense answers for easier reading by adding "my child" to the question. - Prioritise "using a seatbelt" as the first answer	b) Not applicable – we never travel with my child's wheelchair c) My child's wheelchair is folded and placed in the boot d) We have an adapted vehicle with a lift so that my child remains in their wheelchair for the duration of the commute How does your child sit in the vehicle? a) They sit on the seat, on another passenger's lap or lie down and DO NOT use a seatbelt b) They sit on the seat and USE a seatbelt c) They stay in their wheelchair and their wheelchair is secured within the vehicle d) They sit in a car seat and are within the recommended height and weight e) They sit in a car seat or positioner designed for children with physical disabilities such as the ShonaQuip vehicle positioner or the BeSafe IziUp chair. f) I must adapt the current setup with additional pillows, blankets or a home-made device to help my child be more comfortable	b) ~~Not applicable – w~~[We] never travel with my child's wheelchair [because ______________] c) My child's wheelchair is folded and placed in the ~~boot~~[vehicle] d) We have an adapted vehicle with a lift so that my child remains in their wheelchair for the duration of the ~~commute~~[journey] How does your child sit in the vehicle? [My child... a) Sits on the seat and USES a seatbelt] b) ~~They s~~[Sits] on the seat, on another passenger's lap or lie down ~~and~~ [but] DO NOT use a seatbelt ~~They sit on the seat and USE a seatbelt~~ c) ~~They s~~[Stay] in their wheelchair ~~and their wheelchair~~ [which] is secured ~~within~~ [to] the vehicle [floor] d) ~~They s~~[Sits] in a [normal] car seat ~~and are within the recommended height and weight~~ [or booster seat] e) ~~They s~~[Sits] in a car seat or positioner designed for children with physical disabilities such as the ShonaQuip vehicle positioner or the BeSafe IziUp chair. f) ~~I must~~ [Is not comfortable unless I] adapt the current setup with additional pillows,
Question 13		

- Divide this question to adequately describe the challenges with either public or private transport. - Further describe the challenge faced when putting a child into the vehicle - Avoid medical jargon and replace "trunk" with "chest"	Do you have any difficulties when travelling with your child in a vehicle? If so, please tick any of the relevant boxes. You may tick more than one if necessary. ☐ No space next to car to place wheelchair close to car ☐ Wheelchair/walking frame/assistive device is too bulky to fit into car ☐ Difficulty putting child into vehicle because door is too small ☐ Child is unable to fasten seatbelt/safety harness themselves ☐ The seats are too close together and there is not enough space for my child's legs ☐ My child's head or trunk falls over when we turn corners or drive for long distances Other: (Please describe:)________________________	blankets or a home-made device ~~to help my child be more comfortable~~ Do you have any difficulties when travelling with your child in a **private** vehicle? If so, please tick any of the relevant boxes. You may tick more than one if necessary. ☐ No space next to car to place wheelchair close to car ☐ [There is no dedicated Disabled Parking space or it is used by another vehicle] ☐ Wheelchair/walking frame/assistive device/car seat is too bulky to fit into car ☐ Difficulty putting child into vehicle because door is too small ☐ [The seats are too close together and there is not enough space for my child's legs] ☐ My child's head or ~~trunk~~ [chest] falls over when we turn corners or drive for long distances ☐ [My child's weight makes getting into the vehicle difficult] Other: (Please describe:)____________________ Do you have any difficulties when travelling with your child on **public transport?** [(This includes on the school bus)] If so, please tick any of the relevant boxes. You may tick more than one if necessary.

		☐ [There is not enough time to help my child onto the vehicle before the driver leaves] ☐ [The driver refuses to transport my child because they are disabled] ☐ Wheelchair/walking frame/assistive device is too bulky to fit into car ☐ Difficulty putting child into vehicle because door is too small too/too high from the pavement ☐ [No space on the taxi/bus/vehicle for my child and their wheelchair] ☐ The seats are too close together and there is not enough space for my child's legs ☐ [My child cannot sit properly and safely on the seat provided in the vehicle] ☐ [My child's weight makes getting into the vehicle difficult] Other: (Please describe:)________________ [Only answer this question if you indicated "My child **does not go out** because of the challenges faced" in either question 9 or 10.] Please tick which box(es) describe the challenges faced: ☐ [It takes too long to get my child ready for a journey] ☐ [I cannot take my child's wheelchair with us in the vehicle]

		☐ [I feel that it is unsafe to travel with my child] ☐ [My child and their wheelchair cannot be accommodated n the public transport] ☐ [I cannot afford the costs of a private taxi] ☐ [It is too difficult to get my child into the vehicle] ☐ [The public transport is too far from my house/destination and I can't push the wheelchair that far] Other (Please describe:)________________
Question 14 - Split question to allow for differentiation in public and private transport - Condense answers for easier reading by adding "my child" to the question. - Split option c) in safety and no comfort and no safety or comfort	Do you feel that your child has sufficient support to sit comfortably for the duration of the journey? a) I feel that my child sits comfortably for the duration of the journey and they are safety restrained. b) I feel that my child sits comfortably for the duration of the journey but they are NOT safety restrained. c) I feel that my child DOES NOT sit comfortably for the duration of the journey and they are NOT safety restrained.	[When travelling with private or public transport, how] ~~D~~[do] you feel that your child ~~has sufficient support to~~ [sits] ~~comfortably for the duration of~~[throughout] the journey? [My child...] a) ~~I feel that my child s~~[S]its comfortably for the duration of the journey and they are safety restrained. b) ~~I feel that my child s~~[S]its comfortably for the duration of the journey but they are NOT safety restrained. c) ~~I feel that my child~~ DOES NOT sit comfortably for the duration of the journey [but they are safely restrained.] d) [DOES NOT sit comfortably for the duration of the journey] and they are NOT safety restrained.
Question 15 - Improve question as terminology		

creates ambiguity with a seatbelt and a car seat or booster seat	Have you ever used a car restraint system/car seat? a) Yes b) No	~~Have~~ [Has] you[r child] ever used a ~~car restraint system/~~car seat? a) Yes b) No
Question 15.1 - The answers make assumptions about the use of car seats instead of gathering new information.	If you answered yes to Question 15 please indicate why you have stopped using the car seat a) We still use the car seat b) My child outgrew the car seat (too tall or too big) c) My child's disability prevents us from using the car seat anymore	If you answered yes to Question 15 please ~~indicate why you have stopped using the car seat~~[elaborate] a) ~~We~~ [My child] still use[s] ~~the~~ car seat b) My child [no longer uses a car seat because they] outgrew the car seat (too tall or too big) c) My child's disability prevents us from using the car seat anymore
Question 15.2 - Options do not include the parent's intent to use a car seat but no facilities on public transport were available.	If you answered no to Question 15 please indicate why you have never used a car seat? a) I do not think my child needs one b) I cannot afford it c) My child's disability prevents us from using a car seat	If you answered no to Question 15 please indicate why you have never used a car seat? a) I do not think my child needs one b) I cannot afford it c) [There are no facilities on public transport to use a car seat] d) My child's disability prevents us from using a car seat
Legislation - Include an additional question to understand parent's		[According to your understanding, which groups of people should use a seatbelt?

perspective on seatbelt legislation		a) No-one, especially if we drive slowly or do not go on highways b) Only the driver c) The driver and the passenger d) All passengers, even those in the backseat]
Social circumstances Question 19 - Rephrase answers to improve the tone of the options Include an opportunity to add any other comments that the parents wish to give	How much would you be willing to spend on a car seat that is right for your child? a) None – I do not think my child needs a car seat b) None – I cannot afford to spend any money on a car seat c) None – A car seat MUST be provided by Government or my Medical Aid d) Less than R1000 e) R1000 – R3000 f) R3000 – R10 000 g) More than R10 000	How much would you be willing to spend on a car seat that is right for your child? a) ~~None~~ I do not think my child needs a car seat b) None – I cannot afford to spend any money on a car seat c) ~~None – A car seat MUST~~ [I should not have to pay for a car seat, it should] be provided by [the] government or my Medical Aid d) Less than R1000 e) R1000 – R3000 f) R3000 – R10 000 g) More than R10 000 [If you have any further comments about the how your child is transported or the challenged faced, please let us know: ____________________________]

Table 22: Amendments to Self-Designed Transportation Questionnaire during Friends Day Centre focus group discussion

Questionnaire Sub-section and concerns raised during focus group	Original questions	Adjustments made to questions Exclusions shown by strikethrough Inclusions indicated by [...]
Understanding your child's disability Question 4 - Elaborate the question to explain why there is a limit to the suggested diagnoses.	Diagnosis: a) Cerebral Palsy b) Spina Bifida c) Muscular Dystrophy d) Other (please specify) ___________________	[Does your child have a] diagnosis [that may affect their balance]: a) Cerebral Palsy b) Spina Bifida c) Muscular Dystrophy d) Other (please specify) ___________________
For ease of reference a heading was put on top of the page when sub-section continues		**[UNDERSTANDING YOUR CHILD'S DISABILITY CONT.]**
Question 5 - Alter the formatting to emphasise what is being asked by making the phrase bold	Please choose which option best describes your child's ability to sit on a chair. My child...	Please choose which option best describes your child's ability to [**sit on a chair**]. My child...
Question 7 - Alter the formatting to emphasise what is being asked by making the phrase bold	In the last six weeks, has your child undergone any surgery? Please tick box if relevant or list surgery:	In the [**last six weeks**], has your child undergone any surgery? Please tick box if relevant or list surgery:

Transportation Change order of the questions to first relate to travel, then seating and then describe the challenges.		
Question 8 - Alter the formatting to emphasise what is being asked by making the phrase bold	How does your child travel to school?	How does your child travel to [**school**]?
For ease of reference a heading was put on top of the page when sub-section continues		[**TRANSPORTATION CONT.**]
Question 9 - Alter the formatting to emphasise what is being asked by making the phrase bold - Expand the answer to include children who walk with varied assistance or are pushed in a pram	How does your child commute to activities/places near to your house? (less than 5km or 15min drive) a) Walk or pushed in wheelchair	How does your child [usually] commute to [**activities/places near**] to your house? (less than 5km or 15min drive) a) Walk [with/without assistance] or pushed in [pram/] wheelchair
Question 10 - Alter the formatting to emphasise what is being asked by making the phrase bold	How does your child commute to activities/places far from your house? (more than 5km or 15min drive)	How does your child commute to [**activities/places far**] from your house? (more than 5km or 15min drive)

Question 11 - Change formatting to encourage ticking multiple options if applicable - Allow parents the opportunity to describe other challenges not listed	Only answer this question if you indicated "My child does not go out because of the challenges faced" in either question 9 or 10. Please tick which box(es) describe the challenges faced: a. It takes too long to get my child ready for a journey b. I cannot take my child's wheelchair with us in the vehicle c. I feel that it is unsafe to travel with my child d. My child and their wheelchair cannot be accommodated n the public transport e. I cannot afford the costs of a private taxi f. It is too difficult to get my child into the vehicle g. The public transport is too far from my house/destination and I cannot push the wheelchair that far	Only answer this question if you indicated "My child does not go out because of the challenges faced" in either question 9 or 10. Please tick which box(es) describe the challenges faced: □ It takes too long to get my child ready for a journey □ I cannot take my child's wheelchair with us in the vehicle □ I feel that it is unsafe to travel with my child □ My child and their wheelchair cannot be accommodated n the public transport □ I cannot afford the costs of a private taxi □ It is too difficult to get my child into the vehicle □ The public transport is too far from my house/destination and I cannot push the wheelchair that far □ [Other (Please describe:)___________]
Question 12 - Expand the question to include a child's pram	How do you transport your child's wheelchair when commuting?	How do you transport your child's wheelchair[/pram] when commuting?
Question 13		

- Include an option for extended travel as many children commute for long periods of the day	How long does your child spend, on average, in a vehicle per day? a) Less than one hour b) Between one to two hours c) More than two hours d) Not applicable – my child does not travel in a vehicle	How long does your child spend, on average, in a vehicle per day? a) Less than one hour b) Between one to two hours c) [Between two to four hours] d) More than ~~two~~ [four] hours e) Not applicable – my child does not travel in a vehicle
Question 15 - Parents and bus assistants felt that a child fastening themselves independently was not a priority challenge, but rather an emphasis should be put on the lack of disabled parking	Do any of the relevant boxes. You may tick more than one if necessary. ☐ No space next to car to place wheelchair close to car ☐ Wheelchair/walking frame/assistive device is too bulky to fit into car ☐ Difficulty putting child into vehicle because door is too small ☐ Child is unable to fasten seatbelt/safety harness themselves ☐ The seats are too close together and there is not enough space for my child's legs ☐ My child's head or chest falls over when we turn corners or drive for long distances ☐ Other: (Please describe:)_______________	Do any of the relevant boxes. You may tick more than one if necessary. ☐ No space next to car to place wheelchair close to car ☐ [Dedicated Disabled Parking space is used by another vehicle] ☐ Wheelchair/walking frame/assistive device is too bulky to fit into car ☐ Difficulty putting child into vehicle because door is too small ☐ ~~Child is unable to fasten seatbelt/safety harness themselves~~ ☐ The seats are too close together and there is not enough space for my child's legs ☐ My child's head or chest falls over when we turn corners or drive for long distances ☐ Other: (Please describe:)_______________

Question 16 - Parents and bus assistants felt that a child fastening themselves independently was not a priority. Reduce the length of questionnaire by removing answer	Do you have any difficulties when travelling with your child on public transport? If so, please tick any of the relevant boxes. You may tick more than one if necessary. ☐ There is not enough time to help my child onto the vehicle before the driver leaves ☐ Wheelchair/walking frame/assistive device is too bulky to fit into car ☐ Difficulty putting child into vehicle because door is too small too/too high from the pavement ☐ Child is unable to fasten seatbelt/safety harness themselves ☐ No space on the taxi/bus/vehicle for my child and their wheelchair ☐ The seats are too close together and there is not enough space for my child's legs ☐ My child's head or trunk falls over when we turn corners or drive for long distances Other: (Please describe:)________________	Do you have any difficulties when travelling with your child on public transport? If so, please tick any of the relevant boxes. You may tick more than one if necessary. ☐ There is not enough time to help my child onto the vehicle before the driver leaves ☐ Wheelchair/walking frame/assistive device is too bulky to fit into car ☐ Difficulty putting child into vehicle because door is too small too/too high from the pavement ☐ ~~Child is unable to fasten seatbelt/safety harness themselves~~ ☐ No space on the taxi/bus/vehicle for my child and their wheelchair ☐ The seats are too close together and there is not enough space for my child's legs ☐ My child's head or trunk falls over when we turn corners or drive for long distances Other: (Please describe:)________________

		[If your child uses the **school bus (refer to Question 8),** how do you feel that your child sits throughout journey?
Question 17 - Include an additional question for how the child sits during transport to school		My child… a) sits comfortably for the duration of the journey and they are safely restrained. b) sits comfortably for the duration of the journey but they are NOT safely restrained. c) DOES NOT sit comfortably for the duration of the journey but they are safely restrained. d) DOES NOT sit comfortably for the duration of the journey and they are NOT safely restrained.]
Question 18 - Incorrect answer numbering	When travelling with public transport, how do you feel that your child sits throughout journey? My child… e) sits comfortably for the duration of the journey and they are safely restrained. f) sits comfortably for the duration of the journey but they are NOT safely restrained. g) DOES NOT sit comfortably for the duration of the journey but they are safely restrained.	When travelling with public transport, how do you feel that your child sits throughout journey? My child… [a)] sits comfortably for the duration of the journey and they are safely restrained. [b)] sits comfortably for the duration of the journey but they are NOT safely restrained. [c)] DOES NOT sit comfortably for the duration of the journey but they are safely restrained.

Question 15.2 - Incorrect question numbering - Add an opportunity for the parents to explain why or how the disability prevents the use of car seat	h) DOES NOT sit comfortably for the duration of the journey and they are NOT safely restrained. If you answered no to Question 19 please indicate why you have never used a car seat? a) I do not think my child needs one b) I cannot afford it c) There are no facilities on public transport to use a car seat d) My child's disability prevents us from using a car seat	[d)] DOES NOT sit comfortably for the duration of the journey and they are NOT safely restrained. If you answered no to Question 19 please indicate why you have never used a car seat? a) I do not think my child needs one b) I cannot afford it c) There are no facilities on public transport to use a car seat d) My child's disability prevents us from using a car seat [(Please describe:)______________]
Legislation Question 21 - Further description of the question for more clarity Question 22	According to your understanding, which groups of people should use a seatbelt? a) No-one, especially if we drive slowly or do not go on highways b) Only the driver c) The driver and the passenger d) All passengers, even those in the backseat	According to your understanding [of the law], which groups of people should use a seatbelt? a) No-one, especially if we drive slowly or do not go on highways b) Only the driver c) The driver and the passenger d) All passengers, even those in the backseat

| - Further description of the question for more clarity
- Adjust age and height parameters | According to your understanding, which groups of children should use a car seat or booster?
a) None, being held by a passenger on their lap or sitting in the back of the vehicle is okay
b) Only babies under one year
c) Only babies and toddler under 3 years
d) Children under the age of 12 or less than 145cm tall | According to your understanding [of the law], which groups of children should use a car seat or booster?
a) None, being held by a passenger on their lap or sitting in the back of the vehicle is okay
b) Only babies under one year
c) Only babies and toddler under 3 years
Children under the age of ~~12~~ [14] or less than ~~145cm~~ [150cm] tall |
| *Social Circumstances*
No changes/alterations were made | | |

Table 23: Amendments to Self-Designed Transportation Questionnaire during EROS School focus group discussion

Questionnaire Sub-section and concerns raised during focus group	Original questions	Adjustments made to questions Exclusions shown by strikethrough Inclusions indicated by [...]
Understanding your child's disability Question 4 - Emphasise the link between diagnosis and sitting balance	Does your child have a diagnosis that may affect their balance? a) Cerebral Palsy b) Spina Bifida c) Muscular Dystrophy	Does your child have a diagnosis that may affect their [sitting] balance? a) Cerebral Palsy b) Spina Bifida c) Muscular Dystrophy

Question 5	d) Other (please specify) ________________	d) Other (please specify) ________________
Question 5 - Elaborate on the child's ability to sit unattended - Explain where the child requires support during sitting	Please choose which option best describes your child's ability to **sit on a chair**. My child… a) Can sit on any type of chair without using their hands to support them. They can also reach across their body to pick up something without falling over. b) Can sit by themselves on most types of chairs but falls if they reach across the body c) Requires a chair with a backrest or must use their hands for support d) Sits with the assistance of a well-supported chair with a backrest and armrests e) Needs a lot of support from the wheelchair or someone holding them to sit	Please choose which option best describes your child's ability to **sit on a chair** [unattended]. My child… a) Can sit on any type of chair without using their hands to support them. They can also reach across their body to pick up something without falling over. b) Can sit by themselves on most types of chairs but falls if they reach across the body c) Requires a chair with a backrest or must use their hands for support d) Sits with the assistance of a well-supported chair with a backrest and armrests e) Needs a lot of support [for their legs, back and head] from the wheelchair or someone holding them to sit
Transportation Question 12 - Adjust sentence structure - Include an option that describes the challenge of the child's weight	Do you have any difficulties when travelling with your child in a **private vehicle**? If so, please tick any of the relevant boxes. You may tick more than one if necessary.	Do you have any difficulties when travelling with your child in a **private vehicle**? If so, please tick any of the relevant boxes. You may tick more than one if necessary.

	☐ No space next to car to place wheelchair close to car ☐ Dedicated Disabled Parking space is used by another vehicle ☐ Wheelchair/walking frame/assistive device/car seat is too bulky to fit into car ☐ Difficulty putting child into vehicle because door is too small ☐ The seats are too close together and there is not enough space for my child's legs ☐ My child's head or chest falls over when we turn corners or drive for long distances ☐ Other: (Please describe:)_______________	☐ No space next to car to place wheelchair close to car ☐ [There is no] dedicated Disabled Parking space [or it] is used by another vehicle ☐ Wheelchair/walking frame/assistive device/car seat is too bulky to fit into car ☐ Difficulty putting child into vehicle because door is too small ☐ The seats are too close together and there is not enough space for my child's legs ☐ My child's head or chest falls over when we turn corners or drive for long distances ☐ [My child's weight makes getting into the vehicle difficult] ☐ Other: (Please describe:)_______________
Question 13 - Explain that challenges on public transport may include those experienced on the school bus - Include an option that describes the prejudice of the	Do you have any difficulties when travelling with your child on **public transport**? If so, please tick any of the relevant boxes. You may tick more than one if necessary.	Do you have any difficulties when travelling with your child on **public transport**? [(This includes on the school bus)] If so, please tick any of the relevant boxes. You may tick more than one if necessary. ☐ There is not enough time to help my child onto the vehicle before the driver leaves

<table>
<tr><td valign="top">

driver towards the child and one describing the child's weight as a challenge

- Rephrase sentence for ease of understanding

</td><td valign="top">

☐ There is not enough time to help my child onto the vehicle before the driver leaves

☐ Wheelchair/walking frame/assistive device is too bulky to fit into car

☐ Difficulty putting child into vehicle because door is too small too/too high from the pavement

☐ No space on the taxi/bus/vehicle for my child and their wheelchair

☐ The seats are too close together and there is not enough space for my child's legs

☐ My child's head or trunk falls over when we turn corners or drive for long distances

Other: (Please describe:)______________________

</td><td valign="top">

☐ [The driver refuses to transport my child because they are disabled]

☐ Wheelchair/walking frame/assistive device is too bulky to fit into car

☐ Difficulty putting child into vehicle because door is too small too/too high from the pavement

☐ No space on the taxi/bus/vehicle for my child and their wheelchair

☐ The seats are too close together and there is not enough space for my child's legs

☐ My child~~'s head or trunk falls over when we turn corners or drive for long distances~~ [cannot sit properly and safely on the seat provided in the vehicle]

☐ [My child's weight makes getting into the vehicle difficult]

☐ Other: (Please describe:)______________

[Do you feel that your child must be supervised throughout the journey?
    a)   Yes, because______________________
    b)   No]

</td></tr>
<tr><td valign="top">

Include an additional question about the need for supervision as this describes the child's circumstances and behaviours in more detail

</td><td></td><td></td></tr>
</table>

		If you answered yes to Question 20 please elaborate
Question 21 - Add an opportunity for the parents to explain why or how the disability has prevented the use of a car seat	If you answered yes to Question 20 please elaborate a) My child still uses a car seat b) My child no longer uses a car seat because they outgrew the car seat (too tall or too big) c) My child's disability prevents us from using the car seat anymore	a) My child still uses a car seat b) My child no longer uses a car seat because they outgrew the car seat (too tall or too big) c) My child's disability prevents us from using the car seat anymore [(Please describe:)________]
Question 24 - Differentiate between car seat and booster seat	According to your understanding of the law, which groups of children should use a car seat or booster? a) None, being held by a passenger on their lap or sitting in the back of the vehicle is okay b) Only babies under one year c) Only babies and toddler under 3 years d) Children under the age of 14 or less than 150cm tall	According to your understanding of the law, which groups of children should use a car seat or booster [seat]? a) None, being held by a passenger on their lap or sitting in the back of the vehicle is okay b) Only babies under one year c) Only babies and toddler under 3 years d) Children under the age of 14 or less than 150cm tall

UCT Research Study
Physiotherapy Department
Old Main Building
Groote Schuur Hospital
Tel: +27 21 696 4134 / +27 21 406 6059
E-mail: LTTKER002@myuct.ac.za

Today's Date: _______________________________

Please help us to understand more about your child and the different ways they are transported each day. Please fill in any blank spaces or circle the most applicable answer.

UNDERSTANDING YOUR CHILD'S DISABILITY

Child's name: _______________________________________

Child's date of birth: (dd-mm-yyyy)

____

1.   Gender:
     a)   Male
     b)   Female

2.   Height: (Heels to top of the head)
     a)   Less than 95cm
     b)   Between 95- 145cm
     c)   More than 145cm

3.   Weight: (Kgs)
     a)   Less than 13kg
     b)   Between 13 -18kg
     c)   Between 18 – 36kg
     d)   More than 36kg

4.   Does your child have a diagnosis that may affect their sitting balance?
     a)   Cerebral Palsy
     b)   Spina Bifida
     c)   Muscular Dystrophy
     d)   Other (please specify)

     ____

____

____

5. Please choose which option best describes your child's ability to **sit on a chair** unattended. My child...
    a) Can sit on any type of chair without using their hands to support them. They can also reach across their body to pick up something without falling over.
    b) Can sit by themselves on most types of chairs but falls if they reach across the body
    c) Requires a chair with a backrest or must use their hands for support
    d) Sits with the assistance of a well-supported chair with a backrest and armrests
    e) Needs a lot of support for their legs, back and head from the wheelchair or someone holding them to sit

6. Please tick the box(es) if you feel your child has any of the following secondary complications:
    ☐ Spasticity in the muscles (stiffness in the muscles of the arms and/or legs, but they can still be straightened)
    ☐ Joint contractures (Stiff muscles stop joints from being straightened)
    ☐ Scoliosis (curved spine/back)
    ☐ Hip dysplasia (full or partial dislocation of the hip)
    ☐ Loss of sensation (unable to feel part of the body)

7. In the **last six weeks**, has your child undergone any surgery? Please tick box if relevant or list surgery:
    ☐ None
    ☐ Spinal fusion
    ☐ Muscle tendon lengthening
    ☐ Hip surgery
    ☐ Other: (Please describe:)____

    ____

8. How does your child travel to **school**?
    a) Walk or pushed in wheelchair
    b) School combi (small capacity)
    c) School bus (large capacity – different style seats)
    d) Public transport (Train, Taxi, Travel Club, MyCiti/Golden Arrow bus)
    e) Private transport (Car, Combi, Bakkie)

    If you answered d) or e) in Question 8. Please indicate what type of public transport or vehicle is used: ____

    ____

9. How does your child usually commute to **activities/places near** to your house? (less than 5km or 15min drive)
    a) Walk with/without assistance or pushed in pram/wheelchair
    b) Public transport (Train, Taxi, MyCiti/Golden Arrow bus)
    c) Private transport (Car, Combi, Bakkie)
    d) My child does not go out because of the challenges faced

    If you answered b) or c) in Question 9. Please indicate what type of public transport or vehicle is used: ____

10. How does your child commute to **activities/places far** from your house? (more than 5km or 15min drive)
    a) Public transport (Train, Taxi, MyCiti/Golden Arrow bus)
    b) Private transport (Car, Combi, Bakkie)
    c) My child does not go out because of the challenges faced

    If you answered a) or b) in Question 10. Please indicate what type of public transport or vehicle is used: ____
    ____

11. Do you have any difficulties when travelling with your child in a **private vehicle**? If so, please tick any of the relevant boxes. You may tick more than one if necessary.
    ☐ No space next to car to place wheelchair close to car
    ☐ There is no dedicated Disabled Parking space or it is used by another vehicle
    ☐ Wheelchair/walking frame/assistive device/car seat is too bulky to fit into car
    ☐ Difficulty putting child into vehicle because door is too small
    ☐ The seats are too close together and there is not enough space for my child's legs
    ☐ My child's head or chest falls over when we turn corners or drive for long distances
    ☐ My child's weight makes getting into the vehicle difficult
    ☐ Other: (Please describe:)____

12. Do you have any difficulties when travelling with your child on **public transport**? (This includes on the school bus) If so, please tick any of the relevant boxes. You may tick more than one if necessary.
    ☐ There is not enough time to help my child onto the vehicle before the driver leaves
    ☐ The driver refuses to transport my child because they are disabled
    ☐ Wheelchair/walking frame/assistive device is too bulky to fit into car
    ☐ Difficulty putting child into vehicle because door is too small too/too high from the pavement
    ☐ No space on the taxi/bus/vehicle for my child and their wheelchair
    ☐ The seats are too close together and there is not enough space for my child's legs
    ☐ My child cannot sit properly and safely on the seat provided in the vehicle
    ☐ My child's weight makes getting into the vehicle difficult
    ☐ Other: (Please describe:) ____

13. Only answer this question if you indicated "My child **does not go out** because of the challenges faced" in either question 9 or 10.
    Please tick which box(es) describe the challenges faced:
    ☐ It takes too long to get my child ready for a journey
    ☐ I cannot take my child's wheelchair with us in the vehicle
    ☐ I feel that it is unsafe to travel with my child
    ☐ My child and their wheelchair cannot be accommodated n the public transport
    ☐ I cannot afford the costs of a private taxi
    ☐ It is too difficult to get my child into the vehicle
    ☐ The public transport is too far from my house/destination and I can't push the wheelchair that far
    ☐ Other (Please describe:)____

14. How do you transport your child's wheelchair/pram when commuting?
    a) My child does not have a wheelchair
    b) We never travel with my child's wheelchair because

    ____

    ____

    ____

    c) My child's wheelchair/pram is placed in the vehicle
    d) We have an adapted vehicle with a lift so that my child remains in their wheelchair for the
       duration of the journey

15. How long does your child spend, on average, in a vehicle per day?
    a) Less than one hour
    b) Between one to two hours
    c) Between two to four hours
    d) More than four hours
    e) Not applicable – my child does not travel in a vehicle

16. How does your child sit in the vehicle?
    My child…
    a) Sits on the seat and USES a seatbelt
    b) Sits on the seat, on another passenger's lap or lie down but DO NOT use a seatbelt
    c) Stay in their wheelchair which is secured to the vehicle floor
    d) Sits in a normal car seat or booster seat
    e) Sits in a car seat or positioner designed for children with physical disabilities such as the
       ShonaQuip vehicle positioner or the BeSafe IziUp chair.
    f) Is not comfortable unless I adapt the current setup with additional pillows, blankets or a
       home-made device

17. If your child uses the **school bus (refer to Question 8),** how do you feel that your child sits
    throughout journey?
    My child…
    a) sits comfortably for the duration of the journey and they are safely restrained.
    b) sits comfortably for the duration of the journey but they are NOT safely restrained.
    c) DOES NOT sit comfortably for the duration of the journey but they are safely restrained.
    d) DOES NOT sit comfortably for the duration of the journey and they are NOT safely
       restrained.

18. When travelling with **private transport (refer to Questions 8-10),** how do you feel that your child
    sits throughout journey?
    My child…
    a) sits comfortably for the duration of the journey and they are safely restrained.
    b) sits comfortably for the duration of the journey but they are NOT safely restrained.
    c) DOES NOT sit comfortably for the duration of the journey but they are safely restrained.
    d) DOES NOT sit comfortably for the duration of the journey and they are NOT safely
       restrained.

19. When travelling with **public transport (refer to Questions 8-10)**, how do you feel that your child sits throughout journey? My child...
    a) sits comfortably for the duration of the journey and they are safely restrained.
    b) sits comfortably for the duration of the journey but they are NOT safely restrained.
    c) DOES NOT sit comfortably for the duration of the journey but they are safely restrained.
    d) DOES NOT sit comfortably for the duration of the journey and they are NOT safely restrained.

20. Do you feel that your child must be supervised throughout the journey?
    a) Yes, because ____
    ____
    b) No

21. Has your child ever used a car seat?
    a) Yes
    b) No

22. If you answered yes to Question 21 please elaborate
    a) My child still uses a car seat
    b) My child no longer uses a car seat because they outgrew the car seat (too tall or too big)
    c) My child's disability prevents us from using the car seat anymore (Please describe:)_______________ ____
    ____

23. If you answered no to Question 21 please indicate why you have never used a car seat?
    a) I do not think my child needs one
    b) I cannot afford it
    c) There are no facilities on public transport to use a car seat
    d) My child's disability prevents us from using a car seat (Please describe:)_______________________ ____
    ____

24. According to your understanding of the law, which groups of people should use a seatbelt?
    a) No-one, especially if we drive slowly or do not go on highways
    b) Only the driver
    c) The driver and the passenger
    d) All passengers, even those in the backseat

25. According to your understanding of the law, which groups of children should use a car seat or booster seat?
    a) None, being held by a passenger on their lap or sitting in the back of the vehicle is okay
    b) Only babies under one year
    c) Only babies and toddler under 3 years
    d) Children under the age of 14 or less than 150cm tall

26. What is the household monthly income (salaries, wages and SASSA grants from everyone living at home)?
    a) Less than or equal to R1 600
    b) Between R1 600 – R2 500
    c) Between R2 500 – R6 500
    d) Between R6 500 – R15 600
    e) More than R15 600

27. How many persons does this average monthly income support?
    ____

28. How much would you be willing to spend on a car seat that is right for your child?
    a) I do not think my child needs a car seat
    b) None – I cannot afford to spend any money on a car seat
    c) I should not have to pay for a car seat, it should be provided by the government or my Medical Aid
    d) Less than R1000
    e) R1000 – R3000
    f) R3000 – R10 000
    g) More than R10 000

If you have any further comments about the how your child is transported or the challenges faced, please let us know:

____
____
____
____

Thank you for helping us to understand your child and how they are transported.

7.3   COSMIN Checklist - Questionnaire

Properties that have been assessed in this study are marked with x			
A. Internal consistency			
B. Reliability			
C. Measurement error			
D. Content validity (including face validity)			X
Construct validity			
E. Structural validity			
F. Hypotheses testing			
G. Cross-cultural validity			
H. Criterion validity			
I. Responsiveness			
J. Interpretability			

A. Internal consistency			
Design requirements	Yes	No	NA
1. Was the percentage of missing items given?			
2. Was there a description of how missing items were handled?			
3. Was the sample size included in the analysis adequate?			
Statistical methods			
4. Was the Cronbach's alpha calculated?			

Reliability (including test-retest, intra- and inter-rater reliability)			
Design requirements			
1. Was the percentage of missing items given?			
2. Was there a description of how missing items were handled?			
3. Was the sample size included in the analysis adequate?			
4. Were at least two measures available?			
5. Were the administrations independent?			
6. Was the time interval stated?			
7. Were the patients stable in the interim period for the construct being measured?			
8. Was the time interval appropriate?			
9. Were the test conditions similar for both measurements?			
Statistical methods			
10. Was the intraclass correlation coefficient (ICC) calculated, for continuous scores?			
11. Was the kappa calculated for dichotomous scores?			

B. Measurement error			
Design requirements and checks were the same as for reliability.			
Statistical methods			
1. Were limits of agreement assessed			

D. Content validity (including face validity)			
Design requirements			

1. Was there an assessment of whether all items refer to relevant aspects of the construct being assessed?	X		
E. Structural validity			
Design requirements			
1. Does the scale consists of effective indicators?			
G. Cross-cultural validity			
Design requirements			
1. Were both the original language in which the HR- PRO instrument was developed and the language into which it was translated described?			
2. Were the expertise of the persons translating the measure described?			
3. Did the translators work independently from one another?			
4. Were the items translate backwards and forwards?			
5. Was there adequate description of how the differences between the original and translated were resolved?			
6. Was the translation reviewed by a committee (i.e. original developer?)			
7. Was the HR-PRO instrument pre-tested (cognitive interviews to check for interpretation, cultural relevance and ease of comprehension? Statistical methods			
8. Was confirmatory factor analysis performed?			
9. Was differential item function between language groups assessed?			
H. Criterion validity			
Design requirements			
1. Can the criterion used be considered as a reasonable "gold standard"			
G. Responsiveness			
Design requirements			
1. Was a longitudinal design of at least two measurements used?			
2. Was the time interval stated?			
3. If anything occurred in the interim period was it adequately described?			
4. Was a portion of patients changed (improved or deteriorated)?			
5. Were hypotheses about changes in score formulated a priori?			
Statistical methods			
1. Were design and statistical methods adequate for the hypotheses to be tested?			
Generalisability			
Was the sample for which the HR-PRO was evaluated adequately described in terms of:			
1. Mean or median age with Std Dev and range?			
2. Distribution of sex?			
3. Important disease characteristics			
4. Settings at which the study was conducted?			
5. Language in which the instrument was evaluated?			
7. Was the method used for selection of participants described?			
8. Was the percentage of missing responses acceptable?			

7.4 Human Research Ethics Committee Approval

UNIVERSITY OF CAPE TOWN
Faculty of Health Sciences
Human Research Ethics Committee

Room E53-46 Old Main Building
Groote Schuur Hospital
Observatory 7925
Telephone [021] 406 6492
Email: sumayah.ariefdier@uct.ac.za
Website: www.health.uct.ac.za/fhs/research/humanethics/forms

27 March 2018

HREC REF: 837/2017

Dr L Corten
Division of Physiotherapy
Department of Health & Rehab Sciences
F-45 OMB

Dear Dr Corten

PROJECT TITLE: CROSS-SECTIONAL ANALYSIS OF THE POSTURAL SUPPORT NEEDS DURING TRANSPORTATION OF LEARNERS WITH A PHYSICALLY DISABILITY IN THE WESTERN CAPE (MSc-candidate-KA Phillips)

Thank you for your response, addressing the issues raised by the Human Research Ethics Committee (HREC).

It is a pleasure to inform you that the HREC has **formally approved** the above-mentioned study.

Approval is granted for one year until the 28 March 2019.

Please submit a progress form, using the standardised Annual Report Form if the study continues beyond the approval period. Please submit a Standard Closure form if the study is completed within the approval period.
(Forms can be found on our website: www.health.uct.ac.za/fhs/research/humanethics/forms)

We acknowledge that the student: Ms K Phillips will also be involved in this study.

Please quote the HREC REF in all your correspondence.

Please note that the ongoing ethical conduct of the study remains the responsibility of the principal investigator.

Please note that for all studies approved by the HREC, the principal investigator **must** obtain appropriate institutional approval, where necessary, before the research may occur.

Yours sincerely

Signature Removed

PROFESSOR M BLOCKMAN
CHAIRPERSON, FHS HUMAN RESEARCH ETHICS COMMITTEE

Federal Wide Assurance Number: FWA00001637.
Institutional Review Board (IRB) number: IRB00001938

HREC:837-2017

Directorate: Research

Audrey.wyngaard@westerncape.gov.za
tel: +27 021 467 9272
Fax: 0865902282
Private Bag x9114, Cape Town, 8000
wced.wcape.gov.za

REFERENCE: 20180403–913
ENQUIRIES: Dr A T Wyngaard

Mrs Kerry-Ann Phillips
PO Box 263
Edgemead
7407

Dear Mrs Kerry-Ann Phillips

**RESEARCH PROPOSAL: CROSS-SECTIONAL ANALYSIS OF THE POSTURAL SUPPORT NEEDS
DURING TRANSPORTATION OF LEARNERS WITH A PHYSICALLY DISABILITY IN THE WESTERN CAPE**

Your application to conduct the above-mentioned research in schools in the Western Cape has been approved
subject to the following conditions:
1.   Principals, educators and learners are under no obligation to assist you in your investigation.
2.   Principals, educators, learners and schools should not be identifiable in any way from the results of the
         investigation.
3.   You make all the arrangements concerning your investigation.
4.   Educators' programmes are not to be interrupted.
5.   The Study is to be conducted from **10 April 2018 till 28 September 2018**
6.   No research can be conducted during the fourth term as schools are preparing and finalizing syllabi for
         examinations (October to December).
7.   Should you wish to extend the period of your survey, please contact Dr A.T Wyngaard at the contact
         numbers above quoting the reference number?
8.   A photocopy of this letter is submitted to the principal where the intended research is to be conducted.
9.   Your research will be limited to the list of schools as forwarded to the Western Cape Education
         Department.
10.   A brief summary of the content, findings and recommendations is provided to the Director: Research
         Services.
11.   The Department receives a copy of the completed report/dissertation/thesis addressed to:
                 The Director: Research Services
                 Western Cape Education Department
                 Private Bag X9114
                 CAPE TOWN
                 8000

We wish you success in your research.

Kind regards.
Signed: Dr Audrey T Wyngaard
Directorate: Research
DATE: 06 April 2018

Audrey.wyngaard@westerncape.gov.za
tel: +27 021 467 9272
Fax: 0865902282
Private Bag X9114, Cape Town, 8000
wced.wcape.gov.za

REFERENCE: 20180403–913
ENQUIRIES: Dr A T Wyngaard

Mrs Kerry-Ann Phillips
PO Box 263
Edgemead
7407

Dear Mrs Kerry-Ann Phillips

RESEARCH PROPOSAL: CROSS-SECTIONAL ANALYSIS OF THE POSTURAL SUPPORT NEEDS DURING TRANSPORTATION OF LEARNERS WITH A PHYSICALLY DISABILITY IN THE WESTERN CAPE

Your application to conduct the above-mentioned research in schools in the Western Cape has been approved subject to the following conditions:

1. Principals, educators and learners are under no obligation to assist you in your investigation.
2. Principals, educators, learners and schools should not be identifiable in any way from the results of the investigation.
3. You make all the arrangements concerning your investigation.
4. Educators' programmes are not to be interrupted.
5. The Study is to be conducted from **14 January 2019 till 15 March 2019**
6. No research can be conducted during the fourth term as schools are preparing and finalizing syllabi for examinations (October to December).
7. Should you wish to extend the period of your survey, please contact Dr A.T Wyngaard at the contact numbers above quoting the reference number?
8. A photocopy of this letter is submitted to the principal where the intended research is to be conducted.
9. Your research will be limited to the list of schools as forwarded to the Western Cape Education Department.
10. A brief summary of the content, findings and recommendations is provided to the Director: Research Services.
11. The Department receives a copy of the completed report/dissertation/thesis addressed to:
    The Director: Research Services
    Western Cape Education Department
    Private Bag X9114
    CAPE TOWN
    8000

We wish you success in your research.

Kind regards.
Signed: Dr Audrey T Wyngaard
Directorate: Research
DATE: 16 October 2018

Requesting permission to conduct study at their school.

UCT Research Study
Physiotherapy Department
Old Main Building
Groote Schuur Hospital
Tel: +27 21 696 4134 / +27 21 406 6059
E-mail: LTTKER002@myuct.ac.za

Dear Principal,

My name is Kerry-Ann Phillips and I am a physiotherapist at Bel Porto School. I am also completing my MSc at the University of Cape Town. I am doing a study on the use of car seats for children with physical disabilities. The results of this study will help us to better understand the use of and need for specialised car seats for physically disabled children in the Western Cape.

I request your permission to conduct research at your school. As it is a cross-sectional analysis, there will only be assessments and there will be no intervention for the learners.

Data collection procedure will be as follows:

A focus group discussion consisting on a selection of parents, bus drivers/assistants and therapists will assist to validate a new screening tool.

All parents will be sent information packs containing an explanation of the study, informed consent form and short questionnaire. School therapists will be asked to screen learners on their sitting balance. Learners who show difficulty in sitting balance will receive further balance assessment and a simulation of travelling in a car seat. This entails a quick 5-minute sitting assessment to classify their sitting ability followed by testing of different car seats. The learners will be placed in a car seat that is secured to a wheelchair and pushed along a short course to experience similar movements that occur in a moving vehicle. Photographs will be taken of the learners' posture in the car seat before and after the course.

What does this mean for the school?

Focus group participants will be contacted to arrange a convenient time and venue for the discussion.

The school therapists will be asked to assist in distributing and collecting the informed consent forms and questionnaires from the parents as well as screening the learners on their sitting balance. This screening can be done using prior knowledge of the learner's abilities by the therapist. This is a quick process as it is just a screening, however large number of pupils may increase the required time to complete.

During the assessment of the learner's balance, I require a venue such as a small hall with a therapy bench and a plug point. All other equipment will be brought along. Learners will need to be out of class for up to 30 minutes depending on the assessment. This can be done during their normal therapy session, but I request that I can assess learners immediately after each other for smooth running.

I will require medical information such as date of birth, diagnosis, ability of sitting balance and any secondary complications. This can be obtained either through medical records or verbally from school therapists.

I request that a school appointed para-medical personnel, such as the school nurse, be available with the necessary equipment and a first aid kit in the eventuality there is an incident to ensure the safety during the testing phase.

What does it mean for the learners?

The learners, screened by school therapists, with difficulty sitting will have their balance tested and their posture in different car seats documented. There will be no intervention or treatment of the learners.

The learners have the choice to participate in the study, should they feel uncomfortable or wish to stop they may just ask.

Confidentiality

The information that we collect from this research project will be kept confidential. Any information about the learners will have a number on it instead of his/her name. Only the investigator will know what information belongs to each learner. The photographs of the learners before and after the short course will be used to determine their posture and be kept electronically under password protection. The images will not be distributed. If any image is used in a publication, the face will be blocked out. Participation in this research does not involve extra costs for you. No compensation will be given to the school or learners.

If you have any questions or worries after reading this or at any other time during or after the study, please do not hesitate to contact us. You may contact the University of Cape Town Faculty of Health Sciences Human Research Ethics Committee (HREC) if you have any questions or concerns regarding the learner's rights or welfare as research participants.

Kerry-Ann Phillips
021 696 4134
LTTKER002@myuct.ac.za

Dr Lieselotte Corten
021 406 6059
l.corten@uct.ac.za

Human Research Ethics Committee (HREC)
Prof Marc Blockman, Chair
Old Main Building, Groote Schuur Hospital
021 406 6338

1

7.7   Informed Consent Form – Focus Group Discussion
For the participants in the focus group as part of the pilot study

UCT Research Study
Physiotherapy Department
Old Main Building
Groote Schuur Hospital
Tel: +27 21 696 4134 / +27 21 406 6059
E-mail: LTTKER002@myuct.ac.za

This Informed Consent Form has two parts:

- Part I: Information Sheet (to share information about the study with you)
- Part II: Certificate of Consent (for signatures if you agree that you will assist with this study)

You will be given a copy of the full Informed Consent Form

PART I: Information sheet

I, Kerry-Ann Phillips, a qualified Physiotherapist and currently a MSc- student at the University of Cape Town, am doing research on the transportation of physically disabled learners in the Western Cape.

What is the reason for the study and how will it be done?
New laws have been passed on the use of car seats for children during transportation in motor vehicles. These laws use criteria such as age, weight and height of the child to determine if they should be seated in a car seat. Physical disability can make it more challenging to use standard car seats effectively and safely. We would like to determine if the children are currently being transported safely and effectively. We have invited all the learners at your school to take part in this research because they are between the ages of 4 and 18 years.

What does the study mean for you?
A focus group discussion comprising of a selection of parents, bus drivers/class assistants and therapists will happen at the school. You will be asked to meet at the school, where you will meet the other participants in the focus group discussion. You will be led by myself in reading and critiquing the proposed questionnaire on learner transport that will later be sent to parents. The aim is to ensure that the most important questions regarding transport are asked in logical order and a respectful manner.

Voluntary participation
Your decision to participate in this study is entirely voluntary. If you choose not to consent, nothing will change.

Confidentiality
The information that we collect from this research project will be kept confidential. Due to the nature of a focus group discussion, information shared with other members will be public. Participants are requested to remain mindful of others' opinions and views. Participation in this research does not involve extra costs for you. No compensation will be given.

Contact

If you have any questions or worries after reading this form or at any other time during or after the study, please do not hesitate to contact us. You may contact the University of Cape Town Faculty of Health Sciences Human Research Ethics Committee (HREC) if you have any questions or concerns regarding your child's rights or welfare as research participants.

Kerry-Ann Phillips
021 696 4134
LTTKER002@myuct.ac.za

Dr Lieselotte Corten
021 406 6059
l.corten@uct.ac.za

Human Research Ethics Committee (HREC)
Prof Marc Blockman, Chair
Old Main Building, Groote Schuur Hospital
021 406 6338
Marc.Blockman@uct.ac.za

PART II: Certificate of consent

If you consent, we ask you to sign this letter.

I have read the above information, or it has been read to me. I have had the opportunity to ask questions about it and any questions that I have asked have been answered to my satisfaction. I consent voluntarily to participate in this study.

Print Name of Therapist_____________________

Signature of Therapist _____________________

Date _______________________________

For the parent(s)/guardian(s) of children between the ages of 4 and 18 years, enrolled in one of the participating schools in Cape Town.

UCT Research Study
Physiotherapy Department
Old Main Building
Groote Schuur Hospital
Tel: +27 21 696 4134 / +27 21 406 6059
E-mail: LTTKER002@myuct.ac.za

This Informed Consent Form has two parts:

- Part I: Information Sheet (to share information about the study with you)

- Part II: Certificate of Consent (for signatures if you agree that your child may participate)

You will be given a copy of the full Informed Consent Form

PART I: Information sheet

I, Kerry-Ann Phillips, a qualified Physiotherapist and currently a MSc- student at the University of Cape Town, am doing research on the transportation of physically disabled learners in the Western Cape.

What is the reason for the study and how will it be done?

New laws have been passed on the use of car seats for children during transportation in motor vehicles. These laws use criteria such as age, weight and height of the child to determine if they should be seated in a car seat. Physical disabilities can make it more challenging to use standard car seats effectively and safely. We would like to determine if the children are currently being transported safely and effectively. We invite your child to take part in this research because he/she is currently enrolled at one of our participating schools and they are between the ages of 4 and 18 years.

What does the study mean for your child?

We will ask you to fill in a questionnaire concerning your child's disability and how it affects their transportation. This questionnaire will take about 10 minutes to complete.

The children participating in this research will be screened by the school therapists to determine if they have any difficulties with their balance when sitting. Those learners who have difficulty with their balance in sitting will have their balance further tested. This is a simple test of asking your child to sit on a bench with no feet support and determining how well they sit or how much support they need to be able to sit. This will take 5 minutes.

After which, your child's posture in a car seat will be documented through photographs taken before and after a short course. To do this a car seat will be secured to a wheelchair, making it mobile, then your child will be secured within the car seat. The wheelchair will then be pushed along a short course going over small ramps to simulate traveling in a vehicle. This may be repeated in more than one type of car seat. This may take up to 30 minutes. These tests will happen at school during school hours, either during their usual therapy session or at a minimally disruptive time.

What does this study mean for you?

As parent or legal guardian of your child, you will be asked to complete the questionnaire at the beginning of the study. This information that you give will remain confidential. There may be questions that you feel ask for sensitive information, e.g. household income, which you may not feel comfortable sharing in which case you may leave the question open. All your answers will help us to better understand your child's needs and circumstances.

What are the possible benefits and/or risks to your child?

There is a small risk that your child may get injured when being transferred out of their wheelchair for the balance assessment in the CRS. This risk will however be reduced by ensuring safe handling of all children, ensuring the car restraints are properly fastened, and no reckless pushing during the simulation. We will ensure that your child is comfortable and not experiencing any pain throughout the testing by asking for verbal feedback or looking out for gestures such as a facial grimace. The researcher and research assistant/s are trained and experienced in transferring and handling of children with disabilities. The tests used in this study are not harmful in any way. Therefore, there are no direct benefits to participate in the study.

Voluntary participation

Your decision to have your child participate in this study is entirely voluntary. It is your choice whether to have your child participate or not. If you choose not to consent, all the services you and your child receive at the school will continue and nothing will change.

Confidentiality

The information that we collect from this research project will be kept confidential. Any information about your child will have a number on it instead of his/her name. Only the investigator will know what information belongs to your child. The photographs of your child before and after the short course will be used to determine their posture and be kept electronically under password protection. The images will not be distributed. You will be contacted first to ask permission if any image will be used in publication, you may request that your child's face be blocked out. Participation in this research does not involve extra costs for you. No compensation will be given. If funding is available, children participating in car seat investigation part of the study will be given a small gift such as pencil, toy or snack as a token of appreciation.

Contact

If you have any questions or worries after reading this form or at any other time during or after the study, please do not hesitate to contact us. You may contact the University of Cape Town Faculty of Health Sciences Human Research Ethics Committee (HREC) if you have any questions or concerns regarding your child's rights or welfare as research participants.

Kerry-Ann Phillips
021 696 4134
LTTKER002@myuct.ac.za

Dr Lieselotte Corten
021 406 6059
l.corten@uct.ac.za

Human Research Ethics Committee (HREC)
Prof Marc Blockman, Chair
Old Main Building, Groote Schuur Hospital
021 406 6338
Marc.Blockman@uct.ac.za

PART II: Certificate of consent

I have read the above information, or it has been read to me. I have had the opportunity to ask questions about it and any questions that I have asked have been answered to my satisfaction.

YES / NO   I consent to complete the attached Parent Questionnaire

YES / NO   I consent voluntarily for my child to participate as a participant in this study.

YES / NO   I consent to the publication of the photographs of my child's posture with their face
            blanked out (If you choose not to consent to the publication of the photographs, this will
            not affect your child's participation in the study)

Print Name of Participant_____________________

Print Name of Parent or Guardian________________

Signature of Parent or Guardian ___________________

Date ______________________________

Properties that have been assessed in this study are marked with x	
A. Internal consistency	
B. Reliability	X
C. Measurement error	
D. Content validity	
(including face validity)	
Construct validity	
E. Structural validity	
F. Hypotheses testing	
G. Cross-cultural validity	
H. Criterion validity	
I. Responsiveness	
J. Interpretability	

C. Internal consistency			
Design requirements	Yes	No	NA
1. Was the percentage of missing items given?			
2. Was there a description of how missing items were handled?			
3. Was the sample size included in the analysis adequate?			
Statistical methods			
4. Was the Cronbach's alpha calculated?			

Reliability (including test-retest, intra- and inter-rater reliability)			
Design requirements			
1. Was the percentage of missing items given?			X
2. Was there a description of how missing items were handled?			X
3. Was the sample size included in the analysis adequate?	X		
4. Were at least two measures available?	X		
5. Were the administrations independent?	X		
6. Was the time interval stated?	X		
7. Were the patients stable in the interim period for the construct being measured?	X		
8. Was the time interval appropriate?	X		
9. Were the test conditions similar for both measurements?	X		
Statistical methods			
10. Was the intraclass correlation coefficient (ICC) calculated, for continuous scores?	X		
11. Was the kappa calculated for dichotomous scores?	X		

D. Measurement error			
Design requirements and checks were the same as for reliability.			
Statistical methods			
1. Were limits of agreement assessed			

D. Content validity (including face validity)			
Design requirements			
1. Was there an assessment of whether all items refer to relevant aspects of the construct being assessed?	X		

E. Structural validity			
Design requirements			

1. Does the scale consists of effective indicators?			
G. Cross-cultural validity			
Design requirements			
1. Were both the original language in which the HR- PRO instrument was developed and the language into which it was translated described?			
2. Were the expertise of the persons translating the measure described?			
3. Did the translators work independently from one another?			
4. Were the items translate backwards and forwards?			
5. Was there adequate description of how the differences between the original and translated were resolved?			
6. Was the translation reviewed by a committee (i.e. original developer?)			
7. Was the HR-PRO instrument pre-tested (cognitive interviews to check for interpretation, cultural relevance and ease of comprehension?			
Statistical methods			
8. Was confirmatory factor analysis performed?			
9. Was differential item function between language groups assessed?			
H. Criterion validity			
Design requirements			
1. Can the criterion used be considered as a reasonable "gold standard"			
G. Responsiveness			
Design requirements			
1. Was a longitudinal design of at least two measurements used?			
2. Was the time interval stated?			
3. If anything occurred in the interim period was it adequately described?			
4. Was a portion of patients changed (improved or deteriorated)?			
5. Were hypotheses about changes in score formulated a priori?			
Statistical methods			
1. Were design and statistical methods adequate for the hypotheses to be tested?			
Generalisability			
Was the sample for which the HR-PRO was evaluated adequately described in terms of:			
1. Mean or median age with Std Dev and range?			
2. Distribution of sex?			
3. Important disease characteristics			
4. Settings at which the study was conducted?			
5. Language in which the instrument was evaluated?			
7. Was the method used for selection of participants described?			
8. Was the percentage of missing responses acceptable?			

7.10 Screening Tool and Guide

For school therapists to screen learner's sitting balance

This is a self-designed traffic light screening tool, based on a three-category approach, to be used to screen learners with a physical disability. It determines which learners have difficulties with sitting balance, hence are eligible for inclusion for assessment of car restraint systems in this population.

This screening tool should preferably be used by occupational therapists or physiotherapists who are trained in seating.

This tool consists of three outputs; green, orange and red; representing the learner's ability to sit independently.

 Green represents no difficulties sitting independently on any surface, e.g. normal chair, car seat or floor, or under any circumstance e.g. unstable or moving surface.

 Orange represents some difficulties to sit independently, defined by problems when sitting on different surfaces, unstable when in a car or unable to sustain posture for more than five minutes. The learner may or may not make use of a wheelchair.

 Red represents difficulties with sitting in most circumstances or poor sitting posture. These learners are likely to use a specialised postural support wheelchair.

Example of how to use a class list in Excel to indicate scoring for learners:

Before

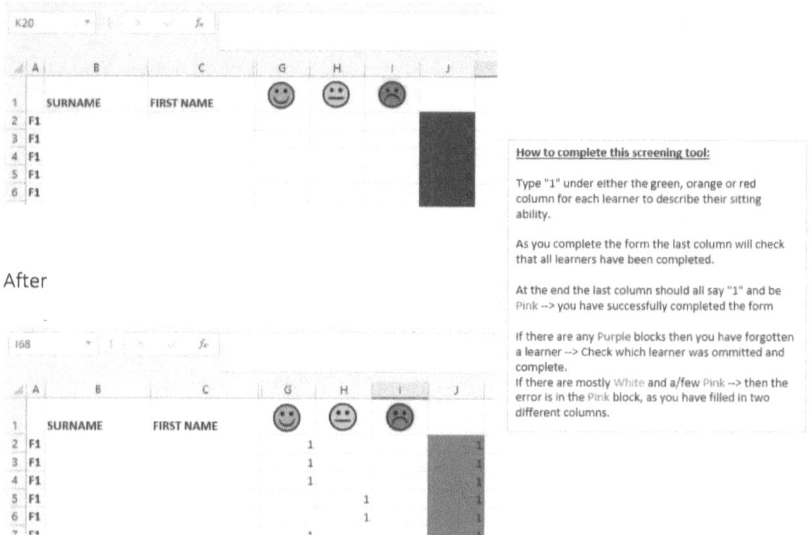

After

7.11 Level of Sitting Scale

Level 1 Unplaceable:	Child cannot maintain the sitting position for 30 seconds while being supported by one person.
Level 2 Supported from Head Downward:	Child requires support of head, trunk and pelvis to maintain the sitting position for 30 seconds.
Level 3 Supported from Shoulders or Trunk Downwards:	Child requires support of trunk and pelvis to maintain the sitting position for 30 seconds.
Level 4 Supported at Pelvis:	Child requires support only at the pelvis to maintain the sitting position for 30 seconds.
Level 5 Maintains Position, Does Not Move:	Child maintains the sitting position independently for 30 seconds, however, movement of any extremity or trunk will result in loss of balance.
Level 6 Shifts Trunk Forward Re-erects:	Child can, without using the hands for support, incline the trunk at least 20 degrees anteriorly in the sagittal plane and return to the neutral (upright) sitting position, independently without losing balance.
Level 7 Shifts Trunk Laterally Re-erects:	Child can, without using the hands for support, incline the trunk at least 20 degrees to one or both sides in the frontal plane and return to the neutral sitting position, independently without losing balance.
Level 8 Shifts Trunk Backwards Re-erects:	Child can, without using the hands for support, incline the trunk at least 20 degrees posterior in the sagittal plane and return to the neutral (upright) sitting position, independently without losing balance.

7.12 Data Collection Form

Today's Date: _________________________   School: _________________________

Learner name	
Date of Birth	
Height (cm)	a) Less than 95cm b) Between 95- 145cm c) More than 145cm
Weight (kg)	a) Less than 13kg b) Between 13 -18kg c) Between 18 – 36kg d) More than 36kg
Diagnosis	
Severity	
Secondary Complications	
Screening Score	Orange / Red

Have you received...	
Parent Informed Consent (signed)	
Learner Assent (signed) *only applicable at EROS	
Parent Questionnaire	

LEVEL OF SITTING SCALE
Level 1 Unplaceable
Level 2 Supported from Head Downward
Level 3 Supported from Shoulders or Trunk Downwards
Level 4 Supported at Pelvis
Level 5 Maintains Position, Does Not Move
Level 6 Shifts Trunk Forward Re-erects
Level 7 Shifts Trunk Laterally Re-erects
Level 8 Shifts Trunk Backwards Re-erects

Today's Date: ________________________   School: __________________________

Learner ID	151
Date of Birth	

SIMULATION	

Car seat type & size	a) Standard Car Seat
	b) Standard Booster Seat
	c) ShonaQuip Vehicle Positioner:
	    I. Small
	    II. Medium
	    III. Large
	d) Be Safe IziUp Car Seat

Why did you choose this car seat?	a) Provides least required support b) Fits anthropometrically
Was it difficult to place the child in the car seat?	a) We experienced no difficulty b) We experienced some difficulty c) We experienced a lot of difficulty d) Unable to place
How well do you think the child is sitting?	a) Child appears comfortable and adequately supported b) Child appears somewhat comfortable and moderately supported c) Child appears uncomfortable and poorly supported
Is the child comfortable and satisfied in the car seat?	a) Child is mostly satisfied with the car restraint b) Child is somewhat satisfied with the car restraint c) Child is somewhat dissatisfied with the car restraint d) Child is mostly dissatisfied with the car restraint
Have you taken a "before" image? Give image number on camera	
Duration of simulation (sec)	
Did you experience any problems during the course?	a) No problems experienced b) Child expressed discomfort; we could remedy it c) Child expressed discomfort and we were unable to remedy it, and stopped simulation d) Child was too uncomfortable to proceed with course
Have you taken an "after" image? Give image number on camera	
Will you re-test in a different car seat?	Yes No
If yes, why?	

7.13 Assent Form

For children older than 5 years of age enrolled at the public special school for learners with specific learning disabilities identified has have difficulty with their balance in sitting.

UCT Research Study
Physiotherapy Department
Old Main Building
Groote Schuur Hospital
Tel: +27 21 696 4134 / +27 21 406 6059
E-mail: LTTKER002@myuct.ac.za

PART I: Information sheet

My name is Kerry-Ann Phillips and I am a physiotherapist and I am also studying at the University of Cape Town. I am doing an investigation into the use of car seats for children with physical disabilities.

I am trying to find out more about how children with physical disabilities travel in motor cars and what type of car seats they use. You are being asked to join this study because you have difficulty with your balance when sitting and are younger than 18 years.

If you decide you want to be in this study, this is what will happen:
Your balance in sitting will be tested by asking you to sit on a bench, and if you need some help we will help you so that you do not fall. We will take note of how well you sit or how much support you need to sit up. This will only take 5 minutes.

After this we will test different car seats by simulating travelling in a vehicle. We will ask you to sit in the car seat which is secured to a wheelchair so that we can push it around a short course over a ramp. Before and after this course we will take a photograph of your position in the car so that we can see how well you sit in the chair.

Can something bad happen to me?
We want to tell you about some things that might hurt or upset you if you are in the study.

There is a chance that something might go wrong during the testing of your balance or simulation in the car seat such as you falling. We will do everything we can to keep you safe and make you feel comfortable. You will not be very high off the ground and we will make sure we do not push you too fast on the course.

If you feel uncomfortable or experience any pain you can tell the examiner and they will stop the test immediately. The tests we will do, will not do anything bad to you.

Can anything good happen to me?
We are just trying to understand how children travel in motor vehicles, we will not be doing any therapy. But we hope to learn something that will help other people someday.

Do I have other choices?
You can choose not to be in this study

Will anyone know I am in the study?

We won't tell anyone you took part in this study. When we are done with the study, we will write a report about what we found out. We won't use your name in the report. The only people who know you are in the study is the investigator and school therapists.

What if I don't want to be in the study?
If you don't want to be in this study, your school therapy will not be affected. You don't have to be in this study. It's up to you. If you say yes now, but you change your mind later, that's okay too. All you have to do is tell us.

Who can I ask questions about the study?
You can always ask questions or tell us about your worries.
The persons you can contact about this are:

Kerry-Ann Phillips   Dr Lieselotte Corten
021 696 4134   021 406 6059
LTTKER002@myuct.ac.za   l.corten@uct.ac.za

PART II: certificate of assent
If you want to be in this study, please sign or print your name.

☐ Yes, I will be in this research study.   ☐ No, I don't want to do this.

_______________________________   _____________________   _____________
Child's name   signature of the child   Date

_______________________________   _____________________   _____________
Person obtaining assent   signature   Date

7.14 Informed Consent Form – School Therapist

For the therapists working at the participating schools in Cape Town.

UCT Research Study
Physiotherapy Department
Old Main Building
Groote Schuur Hospital
Tel: +27 21 696 4134 / +27 21 406 6059
E-mail: LTTKER002@myuct.ac.za

This Informed Consent Form has two parts:

- Part I: Information Sheet (to share information about the study with you)
- Part II: Certificate of Consent (for signatures if you agree that you will assist with this study)

You will be given a copy of the full Informed Consent Form

PART I: Information sheet

I, Kerry-Ann Phillips, a qualified Physiotherapist and currently a MSc- student at the University of Cape Town, am doing research on the transportation of physically disabled learners in the Western Cape.

What is the reason for the study and how will it be done?
New laws have been passed on the use of car seats for children during transportation in motor vehicles. These laws use criteria such as age, weight and height of the child to determine if they should be seated in a car seat. Physical disability can make it more challenging to use standard car seats effectively and safely. We would like to determine if the children are currently being transported safely and effectively. We have invited all the learners at your school to take part in this research because they are between the ages of 4 and 18 years.

What does the study mean for you?
You will be asked to distribute and collect questionnaires that are sent to the parents. These questionnaires concern their child's disability and how it affects their transportation.

The learners participating in this research will be screened by you, the school therapists, to determine if they have any difficulties with their balance when sitting. A self-designed traffic light screening tool, based on a three-category approach, will be used (Appendices 10.5). This tool will consist of three outputs; green, orange and red; representing the learner's ability to sit independently.

Those learners who have difficulty with their balance in sitting will have their balance further tested. After which, their posture in a car seat will be documented through photographs of the learner before and after a simulation. A car seat will be secured to a wheelchair, to make it mobile, and the learner will be secured within the car seat. Then the wheelchair will then be pushed along a short course going over small ramps to simulate traveling in a vehicle. This may be repeated in more than one type of car seat. This may take up to 30 minutes. These tests will happen at school during school hours, either during their usual therapy session or at a minimally disruptive time. You will be asked to help co-ordinate with the educators for an appropriate time for each learner to be seen.

Voluntary participation

Your decision to participate in this study is entirely voluntary. If you choose not to consent, nothing will change.

Confidentiality

The information that we collect from this research project will be kept confidential. Any information about the learners will have a number on it instead of his/her name. Only the investigator will know what information belongs to each child. The photographs of the learners before and after the simulation will be used to determine their posture and be kept electronically under password protection. The images will not be distributed. If any image is used in a publication, the face will be blocked out. Participation in this research does not involve extra costs for you. No compensation will be given.

Contact

If you have any questions or worries after reading this form or at any other time during or after the study, please do not hesitate to contact us. You may contact the University of Cape Town Faculty of Health Sciences Human Research Ethics Committee (HREC) if you have any questions or concerns regarding your child's rights or welfare as research participants.

Kerry-Ann Phillips
021 696 4134
LTTKER002@myuct.ac.za

Dr Lieselotte Corten
021 406 6059
l.corten@uct.ac.za

Human Research Ethics Committee (HREC)
Prof Marc Blockman, Chair
Old Main Building, Groote Schuur Hospital
021 406 6338
Marc.Blockman@uct.ac.za

PART II: Certificate of consent

If you consent, we ask you to sign this letter.

I have read the above information, or it has been read to me. I have had the opportunity to ask questions about it and any questions that I have asked have been answered to my satisfaction. I consent voluntarily to participate in this study.

Print Name of Therapist______________________

Signature of Therapist _____________________

Date _______________________________

7.15 Informed Consent Form – Research Assistant

For the research assistant/s during the study investigating car restraint systems

UCT Research Study
Physiotherapy Department
Old Main Building
Groote Schuur Hospital
Tel: +27 21 696 4134 / +27 21 406 6059
E-mail: LTTKER002@myuct.ac.za

This Informed Consent Form has two parts:

- Part I: Information Sheet (to share information about the study with you)
- Part II: Certificate of Consent (for signatures if you agree that you will assist with this study)

You will be given a copy of the full Informed Consent Form

PART I: Information sheet

I, Kerry-Ann Phillips, a qualified Physiotherapist and currently a MSc- student at the University of Cape Town, am doing research on the transportation of physically disabled learners in the Western Cape.

What is the reason for the study and how will it be done?

New laws have been passed on the use of car seats for children during transportation in motor vehicles. These laws use criteria such as age, weight and height of the child to determine if they should be seated in a car seat. Physical disability can make it more challenging to use standard car seats effectively and safely. We would like to determine if the children are currently being transported safely and effectively. We have invited all the learners at your school to take part in this research because they are between the ages of 4 and 18 years.

What does the study mean for you?

The learners participating in this research will be screened by school therapists to determine if they have any difficulties with their balance when sitting. You will be asked to assist in further testing of those learners who have difficulty with their sitting balance. This will involve transferring learners from their wheelchairs into the car seat, ensure that they are securely fastened and pushing the wheelchair along the short course. During which I will document their posture in a car seat through photographs of the learner before and after a simulation.

A car seat will be secured to a wheelchair, to make it mobile, and the learner will be secured within the car seat. Then the wheelchair will then be pushed along a short course going over small ramps to simulate traveling in a vehicle. This may be repeated in more than one type of car seat. This may take up to 30 minutes. These tests will happen at school during school hours, either during their usual therapy session or at a minimally disruptive time.

Voluntary participation

Your decision to participate in this study is entirely voluntary. If you choose not to consent, nothing will change.

Confidentiality

The information that we collect from this research project will be kept confidential. Any information about the learners will have a number on it instead of his/her name. Only the investigator will know what information belongs to each child. The photographs of the learners before and after the simulation will be used to determine their posture and be kept electronically under password protection. The images will not be distributed. If any image is used in a publication, the face will be blocked out. Participation in this research does not involve extra costs for you. No compensation will be given.

Contact

If you have any questions or worries after reading this form or at any other time during or after the study, please do not hesitate to contact us. You may contact the University of Cape Town Faculty of Health Sciences Human Research Ethics Committee (HREC) if you have any questions or concerns regarding your child's rights or welfare as research participants.

Kerry-Ann Phillips
021 696 4134
LTTKER002@myuct.ac.za

Dr Lieselotte Corten
021 406 6059
l.corten@uct.ac.za

Human Research Ethics Committee (HREC)
Prof Marc Blockman, Chair
Old Main Building, Groote Schuur Hospital
021 406 6338
Marc.Blockman@uct.ac.za

PART II: Certificate of consent

If you consent, we ask you to sign this letter.

I have read the above information, or it has been read to me. I have had the opportunity to ask questions about it and any questions that I have asked have been answered to my satisfaction. I consent voluntarily to participate in this study.

Print Name of Therapist____________________

Signature of Therapist ____________________

Date ___________________________

7.16 Brochure for Parents

UCT Research Study
Physiotherapy Department
Old Main Building
Groote Schuur Hospital
Tel: +27 21 696 4134 / +27 21 406 6059
E-mail: LTTKER002@myuct.ac.za

Thank you for participating in our research study on the transportation of physically disabled learners in the Western Cape by completing a questionnaire.

We really appreciate your feedback. Here are some resources for organisations that are promoting road safety for children in our country.

For more information on the Road Traffic Act on Car Seats; read this article online published by Arrive Alive. *https://www.arrivealive.co.za/Car-Seats-for-Kids-and-Road-Safety-in-South-Africa*

The following organisations are also doing their bit to promote education on road safety in South Africa:

Website: http://drivemoresafely.co.za/
Phone: 081 311 3223
Email: info@drivemoresafely.co.za

Website: http://www.childsafe.org.za
Phone: 021 685 5208
Email: capfsa@pgwc.gov.za

Website: http://www.wheelwell.co.za/
Phone: 073 393 7356 / 072 385 7121
Email: thabile@wheelwell.co.za / peggie@wheelwell.co.za
(As part of their Road Safety initiative Wheelwell tries to assist low-income families with car seats)

Wolrd Health Organisation, *Seat-belts and child restraints: a road safety manual for decision-makers and practitioners*, in *Module 1: The need for seat-belts and child restraints*. 2009: London. p. 6-11

1.2 How seat-belts and child restraints prevent or minimize injury

This section describes what happens during a motor vehicle crash and how seat-belts and child restraints prevent or reduce the severity of injuries sustained.

1.2.1 What happens in a crash?

When a crash occurs, a car occupant without a seat-belt will continue to move at the same speed at which the vehicle was travelling before the collision and will be catapulted forward into the structure of the vehicle – most likely into the steering wheel if they are driving, or into the back of the front seats if they are rear seat passengers. Alternatively, they can be ejected from the vehicle completely. Being ejected from a vehicle drastically increases the probability of sustaining severe serious personal injury or being killed (7).

NOTE Seat-belts as a protection against ejection

The American College of Emergency Physicians advocates the use of seat-belts as the best protection against ejection in a crash. Ejection from a vehicle is one of the most injurious events that can happen to a person in a crash, with 75% of all vehicle occupants ejected from a vehicle in a crash dying as a result. Seat-belts are effective in preventing ejections: overall, 44% of unrestrained passenger vehicle occupants killed are ejected, partially or totally, from the vehicle, as compared to only 5% of restrained occupants (8, 9).

The use of seat-belts and child restraints is one of the most important actions that can be taken to prevent injury in a motor vehicle crash. While seat-belts and child restraints do not prevent crashes from taking place, they play a major role in reducing the severity of injury to vehicle occupants involved in a collision. An occupant's chance of survival increases dramatically when appropriately restrained.

1.2.2 How a seat-belt works

Seat-belts and child restraints are secondary safety devices and are primarily designed to prevent or minimize injury to a vehicle occupant when a crash has occurred. Seat-belts and child restraints thus:

- reduce the risk of contact with the interior of the vehicle or reduce the severity of injuries if this occurs;
- distribute the forces of a crash over the strongest parts of the human body;
- prevent the occupant from being ejected from the vehicle at an impact;
- prevent injury to other occupants (for example in a frontal crash, unbelted rear-seated passengers can be catapulted forward and hit other occupants).

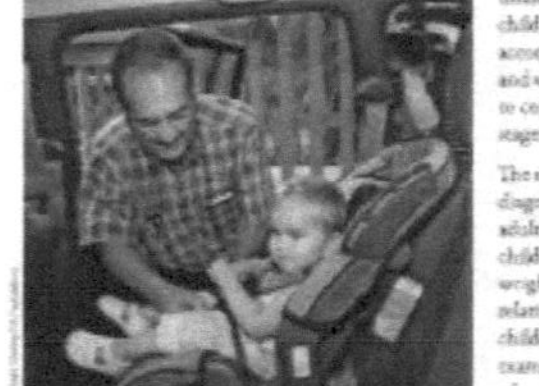

A belted occupant will be kept in their seat and thus will reduce speed at the same rate as the car, so that the mechanical energy to which the body is exposed will be greatly reduced.

1.2.3 How a child restraint works

Infants and children need a child restraint system that accommodates their size and weight, and can adapt to cope with the different stages of their development.

The three-point lap and diagonal seat-belt used by adults is not designed for children's varying sizes, weights, and the different relative proportions of children's bodies. For example, a smaller portion of a child's abdomen is

6

7

covered by the pelvis and rib-cage, while a child's ribs are more likely than an adult's to bend rather than break, resulting in energy from a collision being transferred to the heart and lungs (16). Consequently three-point lap and diagonal seat-belts may lead to abdominal injuries among children, and will not be optimally effective at preventing ejection and injury among them.

Appropriate child restraint systems are specifically designed to protect infants and young children from injury during a collision or a sudden stop by restraining their movement away from the vehicle structure and distributing the forces of a crash over the strongest parts of the body, with minimum damage to the soft tissues. Child restraints are also effective in reducing injuries that can occur during non-crash events, such as a sudden stop, a swerving evasive manoeuvre or a door opening during vehicle movement (17).

1.3 Recommended types of seat-belts and child restraints

1.3.1 Seat-belt design

This section describes the main elements of seat-belt design. Seat-belt designs should comply with national or international standards (covered in Module 1 of the manual). Designs that ensure ease of use will serve to increase wearing rates.

The three-point lap and diagonal seat-belt is the safest and most commonly used in cars, vans, minibuses, trucks and the driver's seat of buses and coaches, while the two-point lap belt is most commonly used in buses and coaches. Seat-belt standards set out requirements for the width of webbing and buckles, and the ease of operation and adjustment. In more recent years seat-belts have become integrated into overall vehicle safety systems that include such devices as pretensioners, load limiters and airbags.

Three-point lap and diagonal seat-belt

Rated highly for effectiveness and ease of use, the three-point lap and diagonal seat-belt is the most commonly used in cars, vans, minibuses and trucks and in the driver's seat of buses and coaches. The seat-belt tongue clips into the buckle, which in the front seats of cars is usually placed on the end of a stiff stalk or directly attached to the seat. A retractor device is included as part of the belt system, as this ensures unnecessary slack is taken up automatically. This system allows the occupant to connect the tongue and buckle using one hand, preventing ejection after maintaining the seating position of the occupant.

Two-point lap belt

A two-point lap belt (sometimes called a "single lap belt") using a retractor device is inferior to the three-point lap and diagonal seat-belt described above but can be sufficient to maintain the seating position of the occupant, particularly in coaches or buses.

Crash studies have shown that although the lap belt does fulfil the task of reducing ejection, it fails to prevent the occupant's head and upper body moving forward and hitting the vehicle interior. For the driver, this could result in serious head injuries from contact with the steering wheel. However, because of the size and mass of coaches, the severity of injury when involved in a collision with another vehicle is often minor compared to that other vehicle if it is a car or van.

Single diagonal belt

The single diagonal design does provide better restraint for the upper body of the wearer than the two-point lap belt, but has been shown to be poorer at preventing ejection and submarining (slipping under the seat-belt).

Full harness

The full harness (double shoulder, lap and thigh straps with central buckle device) gives very good protection, both from ejection and from interior contact. However, it is somewhat cumbersome to put on, and cannot be easily operated with one hand. This is an important factor in achieving a high wearing rate, and thus the harness only tends to be installed in vehicles used for motor sport, where drivers and co-drivers are at high risk.

1.3.2 Types of child restraints

The safest place for children aged 12 years and under is in the back seat, properly restrained in an approved child safety seat. Specially manufactured child restraints should be used for children. There are a number of different types of restraints. The main determining factor for choice of a child restraint is the child's weight (Table 1.1). Older children who are above the height and weight specifications for using child restraints require a properly fitting three-point lap and diagonal seat-belt when riding in a vehicle.

Table 1.1 Weight categories of child restraints

Group	Description
0	For children of a mass less than 10 kg
0+	For children of a mass less than 13 kg
1	For children of a mass from 9 kg to 18 kg
2	For children of a mass from 15 kg to 25 kg
3	For children of a mass from 22 kg to 36 kg

Infants under the age of 1 year (Group 0 or 0+)

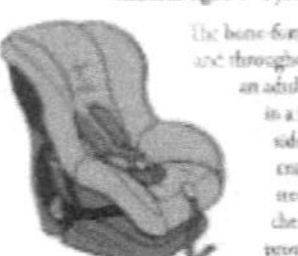

At birth, the infant head is around a quarter of their total length and about a third by weight. An infant's skull is very flexible, so a relatively small impact can cause significant deformation of the skull and brain. The smaller the child, the less force is needed for injury. The infant rib cage is also very flexible. Impact can result in a large compression of the chest wall onto the heart and other abdominal organs. The infant pelvis is unstable and cannot restrain an adult restraint system. Infants require their own special seats to restrain them in a crash, and provide protection from many types of injury. Some seats are convertible, that is, they can revert to a full child safety seat as the child gets older.

A rear-facing child restraint system (sometimes called an "infant car seat") provides the best protection for infants until they are both 1 year of age and at least 13 kilograms (kg) weight. For the best protection, infants should be kept rear facing for as long as possible. The safest place for infants is in the back seat in an approved rear-facing infant car seat.

Children aged 1–4 years (Group I)

The bone-forming process is not complete until the age of 6 or 7 years, and throughout childhood a child's skull remains less strong than that of an adult. A restraint system needs to limit forward head movement in a frontal impact and provide protection from intrusion in a side impact. A child restraint should therefore distribute the crash forces over as wide an area as possible. Belts and harnesses need to fit well and be properly positioned as designed by the manufacturer. The restraint system should also provide protection from contact with the vehicle interior in both front and side impacts.

The best type of child restraint for young children is the child safety seat. The integral harness secures the child and spreads the crash forces over a wide area. This seat will last them until either their weight exceeds 18 kg or they grow too tall for the height of the adjustable harness.

Children aged 4–6 years (Group II)

Booster seats are best used only when a child has outgrown a safety seat. They are designed for children weighing from 15 kg to 25 kg. Children should continue to ride in a booster seat until the lap and diagonal belts in the car fit properly, typically when they are approximately 145 centimetres (cm) tall (22). Booster seats raise the seating position of the child so that the adult seat-belt lies properly across the chest, crossing the child's shoulder rather than the neck, and low across the pelvis. If too high across the stomach, in a crash serious internal injury could result. The child could submarine under the seat-belt. The booster seat has a back that gives some protection in a side impact.

5–11 years (Group III)

Booster cushions without backs are designed for weights from 22 kg to 36 kg, but manufacturers are now producing booster cushions with backs that cover the full 15 kg to 36 kg range. Shield booster seats, which have a plastic shield in front of the child, offer less protection and should not be used. Booster seats for children aged 4–7 years have been shown to reduce injury risk by 59% compared to seat-belts alone (13).

Recent research suggests that children whose restraints are placed in the centre rear seating position incur less injuries than those placed on the outer seats, although this is in contrast to some earlier research that found that the centre seat was not a good position (14, 15). It should also be noted that although children are best secured in age-appropriate child restraints, if such restraints are not available, it is better to use an adult seat-belt on the child than leave the child unrestrained on the back seat (16, 17).

restraint systems

Currently, most child restraint systems are designed to be installed using the vehicle's seat-belt. ISOFIX is a system that uses purpose-designed mounting points provided in the vehicle to attach the child restraint with a rigid mechanism, rather than using the seat-belt to secure the restraint (18). ISOFIX is increasingly used in Australia and in Europe, and similar systems have been adopted in the United States (LATCH) and in Canada (UAS).